WITH A GOOD VEIL, SMOKER AND GLOVES, YOU MAY DEFY
THE WORKER'S STING

THE
BIGGLE BEE BOOK

A SWARM OF FACTS ON PRACTICAL BEE-
KEEPING, CAREFULLY HIVED

BY

JACOB BIGGLE

ILLUSTRATED

*So work the honey bees, creatures that by a rule in nature teach
the art of order to a peopled kingdom.*—SHAKESPEARE.

Skyhorse Publishing

Skyhorse Publishing books may be purchased in bulk at special discounts for sales promotion, corporate gifts, fund-raising, or educational purposes. Special editions can also be created to specifications. For details, contact the Special Sales Department, Skyhorse Publishing, 307 West 36th Street, 11th Floor, New York, NY 10018 or info@skyhorsepublishing.com.

Skyhorse® and Skyhorse Publishing® are registered trademarks of Skyhorse Publishing, Inc.®, a Delaware corporation.

Visit our website at www.skyhorsepublishing.com.

10 9 8 7 6 5 4 3 2 1

Library of Congress Cataloging-in-Publication Data is available on file.

ISBN: 978-1-62636-142-3

Printed in China

PREFACE

As long as I can remember, even as a boy, there were bees kept on our farm. Father, early in his career as a successful farmer, recognized the importance of keeping bees on the place, and this not alone because of the delicious honey they produced, but rather because their presence meant the proper fertilization of the fruit blossoms.

It is not surprising, therefore, that Harriet, my good wife, to say nothing of Tim and Martha, are all deeply interested in the apiary out in the apple orchard.

Honey has become such a necessary part of our daily diet, especially during the fall and winter when griddle-cakes are in order, that we should be lost without it.

Our fifty colonies of bees, though requiring but a small part of our time, have paid us a larger return than any other stock on the place, and we would as soon part with our team of faithful greys, whose steady steps I have frequently followed behind the plow, as to part with our bees.

Not only as a predigested food has honey found a place upon our table, but in the treatment of colds and throat troubles we have found it to be invaluable.

If there are any so-called secrets in the art of beekeeping, they are fully exposed in the following treatise, and it is my earnest desire that the reader may learn to his profit.

Most of the photographs herein reproduced have been taken by Tim, especially for this little volume.

The book is written not only for the professional beekeeper, although he will find within its lids much that will be new to him and much that he has in his experience proved, but also for the farmer and others interested in rural life who are beekeepers on a small scale, and for those who have not as yet become the happy owners of these interesting little people.

It is a mistaken notion to suppose that bees are naturally vindictive, and prone to sting, as the opposite is the truth; for if one will but avoid doing consciously or unconsciously those things which irritate the bees, he will, in the language of my old friend the late Rev. Lorenzo Langstroth, the father of American beekeeping, "Need to have little more fear of the stings of his bees than the horns of his favorite cow, or the heels of his faithful horse."

If one is timid, however, the use of bee veil, gloves and smoker will enable the beginner to avoid being stung at all.

My readers will doubtless be surprised to know that there are localities where bees are not kept, which would bring in splendid returns to those who will keep them; and even on the roofs of city dwellings hives are kept with success, their busy denizens laying tribute upon almost impossible sources for their honey.

My endeavor has been to eliminate from this book all useless material and proved failures in both methods and implements, give to the reader the cream of it all, and outline only methods that are up to date and successful.

                                        JACOB BIGGLE.

Elmwood.

CONTENTS

THE WHOLE FAMILY TAKES AN INTEREST IN THE WELL-KEPT APIARY

Chapter I

REASONS FOR KEEPING BEES

O velvet bee, you're a dusty fellow, you've powdered your legs with gold!—Jean Ingelow.

If for no other reason than to insure the proper fertilization of fruit and other blossoms every farmer, fruit grower or gardener should keep a few bees upon his grounds.

In our habit of regarding beekeeping from the commercial standpoint, and measuring its profit solely by the amount of honey produced, many of us have overlooked the real mission of the honey-bee in life, which is properly to pollenize our blossoms.

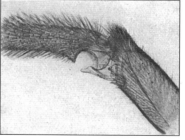

LEG OF WORKER BEE. THE POLLEN IS CARRIED ON THE FUZZY DOWN

In looking at the body of the bee, we discover that it is coated all over with a fuzzy down, to which adheres the pollen of the blossom it enters in quest of nectar, and as the little fellow leaves the blossom with a spiral motion, it unconsciously carries the pollen from the anthers of the blossom to its stigma, and performs an important work which all up-to-date horticulturists have proved.

From an entirely mistaken notion that bees puncture fruit, some fruit growers have been antagonistic to the keeping of bees by their neighbors, and little knew that by so doing they were opposing their own best interests.

Frequently a hornet, wasp or other insect has with its strong mandibles punctured a grape, peach or other fruit, and after taking its fill, has gone its way leaving the sweet juice to continue to exude; then the little bee, coming along and finding the puncture, has taken a sip, and being observed when so doing, has but confirmed the opinion in the mind of the observer that bees are " injurious to fruit."

Bees have no organ sharp or strong enough to accomplish the puncture of fruit, unless it be their sting, and we know absolutely that this organ is never used for that purpose; so bees are surely innocent of the charge brought against them. The late Lorenzo Langstroth, who more than any other man had carefully observed the habits of the honey-bee, once made an experiment that forever silenced all opposers on the subject. He selected a bunch of nice ripe grapes, carefully placed it in a hive of bees, right over the centre of a strong colony, and left it there for several days, and when he removed it he was overjoyed to find that not a grape had been harmed or pierced.

In a number of states, conventions of fruit growers have discussed this question pro and con, and have acknowledged that the keeping of bees was an important factor in the production of fruit. The main damage is done to fruit by birds, and other insects than bees, as proved by the fact that a vineyard or orchard in close proximity to woodland seems to suffer most. The United States Apicultural Station was instructed in 1885 to test the matter thoroughly, by shutting up bees with sound fruit, and the results conclusively showed that the bees were innocent.

Some years ago in California the fruit growers raised such a protest that the beekeepers of that locality were compelled to move their bees from the neighborhood; but when the fruit growers found the following season that their output was greatly reduced, they were glad enough to implore the beekeepers to bring their bees back again, with the result that the output went back to its original quantity.

Up in Schenectady County, N. Y., the heart of the buckwheat country, a man who made it a business to go around in the fall threshing the buckwheat, by machine, declared that when he got into the area that was visited by the bees from one of the largest apiaries in the state, the buckwheat was larger and finer than that of other sections where bees were not in evidence. Yet, strange to say, some of the farmers of that section were inclined to think that the bees robbed the buckwheat of something which the beekeeper had no right to take.

Everyone realizes that in recent years it is a hard matter to get red clover seed that is strong in power of reproduction, and this is due to the fact that the little humblebee has in certain sections of the country become extinct, owing to the breaking up of its habitat in the meadows. The corolla blossom of the red clover is too deep for the ordinary bee to reach the nectar, and the result is that it seldom visits it, and much of it is not properly fertilized.

Until recent years Germany was a large buyer

of our red clover seed, and a few years ago that
government sent to this country a commissioner
to secure thousands of humblebees for introduction
in that country, and some of us recall how the boys
up in New York state were kept busy catching
these little bees for him, with the result that Germany now largely raises its own red clover seed.

The disappearance of the humblebee in some
sections has led the Agricultural Department at
Washington, D. C., to experiment along the line
of increasing the length of the tongues of the
ordinary honey-bees, so that they would visit the
red clover bloom and thus insure its proper fertilization, to say nothing of securing from this
source the millions of pounds of honey that every
year remain unharvested; and the results were so
satisfactory that a large number of queen-bee breeders are selling the well-known "long-tongue red
clover queens," whose bees work on red clover and
secure some of the richest honey and insure a
strong and vigorous seed for the coming season.

There have been cases where fruit growers have
located many miles away from bees, and though
their trees blossomed abundantly, no fruit was
produced until bees were brought into the vicinity.

At Morganville, N. J., Mr. J. F. Becker who has
nearly three acres in glass hothouses, some years
ago conceived the idea that if he could secure the
pollination of cucumber blossoms in the months
of February and March, he could overcome the
climatic advantage of Florida and put into the New
York markets cucumbers that would bring top-
notch prices.

At that season of the year it would have been
folly for him to expect that the bees would be fly-

ing; so to overcome this, he placed in each of his greenhouses a hive of bees, and lured forth by the genial atmosphere they fairly swarmed from blossom to blossom, and as a result enabled him to produce the goods. The bees, however, seemed unable to find their hives again, and died from butting against the glass top of the hothouses, so that it was necessary for him to buy new bees each season; but this was a small item of expense compared with the tremendous crops of cucumbers he annually produced.

BEE FERTILIZING CUCUMBER BLOSSOM

Professor Liberty Bailey of Cornell University declares that "bees are much more efficient agents of pollination than wind, and their absence is always deleterious." Other authorities have experimented by tying a bag around a branch of blossoms until the pollination period had passed, and although the tree bore abundantly on other branches, no fruit was produced by the bloom that had been secluded from the bees.

There are other considerations besides the question of fruit bloom that make beekeeping a delightful adjunct to other farm work—namely, the honey produced and the real pleasure of working among bees.

SOME OF THE MOST SUCCESSFUL BEEKEEPERS ARE WOMEN

CHAPTER II

WHAT RACE TO KEEP

Keep any kind of bees and they'll keep you.—Harriet.

In order to manage bees successfully, it will be well to understand a little of their life history.

The common honey-bee belongs to the order Hymenoptera or membranous winged insects, in the family Apidæ, several types of which compose it—such as Apis or hive-bee, Bombus the humble-bee, and Xylocopa the carpenter-bee, Megachile or leaf-cutter, and Melipoma the stingless-bee of South America.

Our common bee is Apis mellifica, and is scattered all over the country, having been originally imported from Europe in the seventeenth century.

In each hive or colony of bees there are between twenty and forty thousand worker bees, with one queen, and in the spring of the year drones or male bees may be present in large numbers.

The queen bee, easily distinguished by her unusually long tapering body, is the only perfectly developed female in the hive, and thus she is the mother of the family, laying as many as three or four thousand eggs in twenty-four hours and living to be three and four years old. Usually she mates but once, with a drone on the wing, from which union her eggs are fertilized for life. She is in no sense a queen as regards royalty, and there is no evidence that the workers regard her as having royal traits or prerogatives, although some romantic writers have attempted to prove it; on the contrary she is simply regarded as the mother, whose death may result in the extinction of the colony if she can not be replaced, and it is this knowledge that prompts the workers to take such good care of her.

She deposits her eggs one by one in the cells, and in twenty-one days the fully matured worker comes forth. At times she will, by holding her abdomen in a certain position, deposit infertile eggs which produce drone bees, and it usually takes twenty-four days for these eggs to hatch.

Sometimes a colony will become queenless, and

will remedy it by selecting eggs that were intended to produce workers, and, by enlarging their cells and feeding them a stimulative food known as "royal jelly," produce from them other

DEVELOPMENT OF BEE FROM EGG TO BIRTH

queens (in which case they will hatch out in about sixteen days from the time the eggs were laid).

The drones contribute to the warmth of the hive, and are permitted to remain undisturbed until such time as the honey flow or the mating season is over, and are then destroyed or driven from the hive by the little workers. The drones have no stings and gather no honey, their sole purpose in life being to fertilize virgin queens.

The average life of a worker is about five weeks, as they literally work themselves to death; and as for the drones, I suppose they would live some time if the workers did not put them out of business.

One would think that among such a multitude

of bees, everything would be chaos, but just the reverse is the case. The young bees for the first few days of their existence act as nurse bees and feed the unsealed larvæ a mixture of pollen and honey, and do not venture from the hive till several days old, and then only take occasional play-spells flying near the entrance to their home. As they grow older they venture farther from home, and soon become field bees in every sense, gathering nectar, propolis or bee-glue, pollen and water.

There are many races of bees, and their racial characteristics are as pronounced as the dominant traits of different races of the human family. Some are noted for their gentleness, others are more prone to swarming, while others are vicious of temper.

In such a limited treatise it will not be possible to go into an exhaustive account of the various races, but rather will I call attention to races of acknowledged merit, for the average beekeeper is not interested in freaks or costly experiments.

COMMON BLACK OR GERMAN BEES.—These bees need little description, as they are found in almost every locality and are, as their name implies, black. They are much inclined to rob, are very nervous under manipulation, running all over their combs and falling in bunches on the grass. They are easily discour-

QUEEN IN CENTER, DRONE ON LEFT, WORKER AT RIGHT

aged, and more prone to allow the bee-moth to ravage their combs than the yellow races, and have the nasty habit of flying directly at one who approaches their hives. The beekeeper had better pass this race by, and select another that has more desirable traits of character.

CARNIOLANS.—These large grey bees are originally from the cool alpine regions of Carniola, Austria, and are remarkably gentle and the producers of the best comb-honey ever seen. They are

FROM EGG TO FINISHED WORKER

very prolific, rearing much brood, and owing to their hardiness are afield in the morning long before other races, and work later in the afternoon. These bees have many desirable traits; the only fault that can be found with them is that they are great swarmers, but with proper precautions this can be held in check.

ITALIANS.—Perhaps nine-tenths of professional beekeepers keep bees of the Italian race, as they seem to possess, in a larger measure than any others, all desirable traits. They are not only very gentle, but, being of a bright golden hue, are beautiful to behold. This race is very quiet when handled, and will not run all over their combs as do the blacks; and seldom sting unless provoked to it. They are rarely ravaged by the bee-moth, and defend their hives well from robber bees. If properly cared for in the matter of ventilating and shading their hives, they are not much given to swarming.

CAUCASIANS AND BANATS.—These two races of bees are of recent introduction into the country, and are noted for their great gentleness and energy in gathering honey.

There are other races, such as the Cyprians, Holy Lands, Syrians, etc., from the region of the Mediterranean Sea; but so vicious are their tempers that few have dared to keep them, despite the fact that they are great honey gatherers.

Taking all things into consideration, the beginner will make no mistake in adopting the Italian bees, and then as he becomes acquainted with them he can occasionally buy queens of other races and try them in comparison.

The writer has tried many different races and is compelled to say that all things considered the Italian bee is the best general-purpose bee to keep.

A GLASS-SIDED "OBSERVATION HIVE" IS BOTH A
PLEASURE AND A GREAT HELP IN STUDYING
YOUR BEES

CHAPTER III

HOW TO MAKE A START

Hindsight is better than foresight, therefore, start right.
—Farmer Vincent.

There are several ways in which the reader may make a start keeping bees. One way is to buy a few hives of bees from a neighboring beekeeper, or a dealer, and with the aid of new hives make artificial increase by dividing and shaking the bees.

Another way is to purchase bees in old box hives in the fall, and in the spring transfer them to modern hives.

A third way is to buy a complete apiary, a method I should not advise; for, apart from the expense involved, it would be the height of folly to invest a lot of money in such an enterprise, without the requisite experience.

The best way is to start on a small scale and increase your colonies as you become more experienced, and in two or three years you will be able to handle successfully many colonies, whereas at the beginning they would but result in failure. If it is the purpose to make beekeeping a profession, then spend a season or two in the yards of a successful beekeeper; thus the beginner will learn more in one season than by years of unaided work.

By all means get the bees into modern hives as soon as possible, for by so doing disease can be avoided and the work of the apiary carried on expeditiously. The various methods of transferring are treated in the chapter on Spring Manipulation.

The profits of beekeeping will depend upon three things,—the adaptability of the man to the work, the methods pursued, and the number of bees kept in a favorable locality.

While it is true that the question of locality will have an important bearing upon the amount of profit to be derived, yet this can be very easily controlled by not overstocking the neighborhood, and by the use of out apiaries. Although in most sections it would be unwise to keep more than one hundred colonies in each yard, yet by a system of out apiaries, located from three to five miles from the home yard, one can run several hundred colonies, if capital and experience are back of them. (See Chapter IX for details.)

There are of course sections of the country where buckwheat and alfalfa bloom in abundance, allowing several hundred hives to be kept in one yard, but these are not the average conditions. For instance, Mr. E. W. Alexander, of Delanson, N.Y.,

EIGHT HUNDRED HIVES OF BEES IN ONE YARD, THE BEES HAVING ACCESS TO 5,000 ACRES OF BUCKWHEAT

runs nearly eight hundred colonies all in one yard and produces honey by the car-load, his crop one year running up to nearly eighty thousand pounds, principally from buckwheat; but it must be remembered that his location is exceptional. Beekeeping, however, can be made a success in almost any section of the country where agricultural pursuits are successfully carried on, and a fine margin of profit can be derived.

The number of pounds of honey produced by

each colony will depend largely upon the strength of the colony when the flow comes on, and the ability of the beekeeper to have his colonies in good condition for the flow is what means success.

The question of whether comb or extracted honey is to be produced, has very little bearing on the actual profit per colony, for although comb-honey will bring nearly twice as much per pound as the extracted honey, yet a colony will produce nearly double the number of pounds of extracted honey, so that the actual profit per colony will be about the same.

There are some advantages in producing extracted honey (notably that of swarm control), which accounts for the fact that perhaps the majority of professional beekeepers adopt this method, a further discussion of which will be found under the chapter on Comb and Extracted Honey.

Roughly speaking, taking one season with another, there should be an average profit of $4 per colony; this average will often be doubled and in some instances trebled, especially if the honey is sold at retail by the beekeeper.

An experienced beekeeper can without help take care of two hundred colonies, and not be under much expense; but more than this will require an assistant whose services will be needed during the busy season.

Much will depend upon the bloom of the locality where the bees are kept, and if there is much clover and basswood present the returns will often be surprising. As an index to how profitable bees are even under adverse conditions, I find that even in many of our large cities bees are kept on

the roofs of office buildings, by janitors, and give
a fair return for the little attention they require.

However, take my word for it and do not go
heavily into beekeeping, without the experience

BEEKEEPING IS PROFITABLE EVEN
IN A CROWDED CITY

and a location that is
promising, for while
it will give a large re-
turn for the time and
work required, yet it
is not a royal road to
fortune, but requires
intelligent m a n a g e-
ment in many details.
Where the apiary
shall be located is
largely a matter of choice, some placing it in the
orchard, others on the hillside, and others in loca-
tions remote from the house; but wherever it is
placed, be sure to see that it is on well-drained
ground and a safe distance from the live stock, and
within easy reach so that emerging swarms may be
seen and taken care of.

Many beekeepers simply arrange their hives in
a succession of orderly rows, while others have
them in groups of four or five; but though I have
tried almost every method I have never been able
to see that one has any advantage over the other.

See that the grass is well kept and not allowed
to grow so tall as to choke the entrances of hives,
for every such obstruction hinders the bees in their
work; and every such handicap means a loss of
honey. Some beekeepers keep a few sheep in the
bee yard at night; this is a very good way of keeping the grass down, and I have never heard of
either the bees or the sheep suffering thereby.

It is a good practise to have a stand of some kind for each hive, even though it is only a piece of board at the front and the rear of the hive, as this tends to keep the hive out of the mud when it rains, and adds to the life of the bottom board of the hive, to say nothing of the welfare of the colony.

There are some hives that have a combination stand and bottom board, but the majority of bee-keepers use either boards or bricks for the purpose, finding them cheaper and just as effective.

Whatever hive may be adopted, be sure to stick to it as long as it is a success, for it is a nuisance to have different kinds of hives in the same yard; the advantage of having all hive parts interchangeable is manifest, especially during a rush when bees are swarming.

CHAFF HIVES SHADED BY GRAPEVINES

CHAPTER IV

HIVES AND IMPLEMENTS

Not so much depends upon the hive as upon the man behind the hive.—Tim.

The kind of hive to use will be largely determined by whether comb or extracted honey is to be produced. While it is true that here and there a beekeeper will run an apiary for both comb and extracted honey, the majority of them stick to either one or the other.

Whatever the method adopted, by all means do not be led into the folly of attempting to make your own hives.

The writer well recalls some early experiences in home-made hive making, and did not until after some sad experiences learn the wisdom of getting his hives from the supply dealers. The supply dealer with his modern machinery, and the experience of hundreds of beekeepers behind him, is better able to turn out a hive and fixtures than the amateur without that experience.

A VILLAGE ATTEMPT AT A HIVE; IT DOESN'T PAY; BETTER TRANSFER TO A MODERN ONE

When once the beekeeper has used the smooth and well-made hives of some reputable dealer, he will never again return to the poor apologies of his early efforts. Adopt the dovetailed

(27)

hive, as its locked corners are a sure safeguard against leaking or warping, and it will last for many years.

While some beekeepers use the hand-spaced frames, a larger proportion use a self-spacing frame, for by so doing the bees are the more easily moved and the danger of mashing them in going over the colonies is largely overcome.

Much could be said concerning the tops and the bottom boards to be used, but as there are many manufactured, and most of them good, a safe way will be to get the catalogues of a number of manufacturers and take your choice.

If extracted honey is to be the aim, then be sure to adopt the ten-frame hive, as the tendency is more and more toward large hives for extracted honey, for there is no doubt that a booming colony in one hive is far better than two small ones in smaller hives. There are some supply dealers who advocate small eight-frame hives for extracted honey production, and the only reason I can see for their action is that the smaller the hive the more of them must be used.

Use full sheets of foundation in the frames, and not simply starters, and see that they are well wired in horizontally, as this will prevent breaking down of the combs when being whirled around in the extractor.

If the aim is the production of comb-honey, then use a hive having the main body or brood nest more shallow than that to be used for extracted honey, for by so doing the bees are compelled, by lack of storage room over the brood, to carry the surplus up into the surplus section boxes in the upper stories.

A smoker is a necessary adjunct to every apiary, and there are many on the market, some good and some not worth a cent. Personally I prefer the Root Jumbo. The smoker is used for puffing a little smoke in at the entrance of the hive to be opened, which alarms the bees, subdues them, and renders their handling easy. The fuel used varies. Some use dried leaves, others old rags, while others burn planer shavings; I have found the latter to be a very satisfactory fuel. Oil waste is becoming very popular as a fuel.

A GOOD SMOKER WILL CONTROL THE MOST VICIOUS BEES

The idea is to get a fire that will give out lots of smoke and no flame. To accomplish this many beekeepers take old bags, soak them in saltpetre water and allow them to dry, which does the business very satisfactorily. Little pieces of rotten wood, or pieces of dried apple-tree branches, will give a fire that will last for several hours without refilling.

Another necessary article is a bee veil, and it can either be bought of the supply dealer or made at home. Take some black mosquito-netting and sew it to the outside brim of a big straw hat; let it be long enough to tuck down under the coat collar, or else have a drawing string that will permit of its being tied down over the shoulders.

If it is made right not a single bee can ever reach the face.

I never work among the bees except in painter's white cotton suit, for it is with me a well-established fact that bees will not attack so readily if I am clothed in white as when I have on ordinary woolen clothes. Whether this is because they detect the wool of the animal in the woolen suit, or have an aversion to dark colors, I am not able to determine, but I do know that a white cotton suit is best.

A good tool with which to pry off hive lids and elevate frames from the brood nest, is a large

screw-driver, an oyster knife, or a butcher knife; in fact, any one of these will prove to be about as satisfactory as many of the hive tools that are offered for sale.

A good bee dress for women is a pair of men's overalls with drawing string at

A GOOD WHEELBARROW IS A NECESSITY IN THE APIARY

the bottoms, which can be tied about the ankles; and when this has been put on, the skirt can be put over it and no annoyance from the bees will be experienced.

I use a basket in which to carry the smoker, fuel, hive tools and other needful things, and as soon as I am through using one tool I put it back into the basket and carry all from hive to hive; thus they are always at hand when I want them.

I have often been amused to watch a beekeeper hunt in the grass for some tool that has been left around a hive; much valuable time is thus lost. The use of a basket with handle avoids all of this.

SKILL AND FEARLESSNESS COME WITH EXPERIENCE.
SEE THE BEES ON THIS COMB

CHAPTER V

SPRING MANIPULATION

A dead bee gathers no honey.—Martha.

Whether the bees have been wintered indoors or outdoors, it is the height of folly to tinker with them until warm weather is on, as the opening of a hive on cold days allows the warmth of the colony to escape, and may chill and kill much brood, which means a loss of young bees at the honey flow later on when they are most needed.

If the truth were known many a beekeeper has failed to reap a harvest of honey because of a reckless opening of hives during the cold days of early spring.

When settled warm weather is on, the first thing to be done is to go over the apiary and make a careful examination of the colonies, to see if any have died or have lost their queen.

If any have died, remove the hive and its combs to a safe place away from the bees, so that robbing will not be encouraged. If any have lost their queen, remedy the condition at once by giving to that colony a frame of young larvæ or eggs from another colony, being careful not to have the queen of the colony from which it is taken, on it.

A better way is to buy a queen at once, from one of the many reputable queen breeders of the country, and introduce her to the queenless colony according to the directions on the cage in which she comes through the mail. Sometimes a large number of queens will be required if many hives

are kept, and in this case it will pay the beekeeper
to rear his own queens (which see in chapter on
Queen Rearing and Introducing).

Before opening a hive see that the smoker is
smoking satisfactorily, and then go to the hive to
be opened and puff a little smoke in at its entrance
and wait a few moments; then the operation should
be repeated. When the bees appear to be quiet,

OPEN HIVES CAREFULLY; IT WILL SAVE MANY A STING

gently pry up the lid of the hive just a little bit
and send a puff of smoke over the tops of the
frames, and the hive is ready to be opened. It is
a shame the way some beekeepers deluge their
bees with smoke, and then proceed to jerk or
kick the hive lids off. No wonder that in a few
seconds the air is filled with stinging bees, which
could have been entirely avoided had a little more
patience and a little less smoke been used.

After the hive is open, select a frame in the
center of the brood nest and gently lift it from the

hive and examine it carefully to find the queen. I never hunt farther than the first frame if it be well filled with young larvæ, for it is a sign that the queen is present and laying all right.

One by one each hive can thus be examined, and it is astonishing how many hives can be examined in the course of an hour. We have frequently gone over a hundred colonies in four or five hours and not hurried at that.

If any colonies are weak they should be strengthened by giving a frame or two of brood and bees from a strong colony, being careful not to give them the queen; and if suffering from lack of food they should be fed a thick syrup made of equal parts hot water and good granulated sugar; made without cooking.

For this purpose there are many feeders, but nothing is better than an ordinary Mason glass fruit-jar with a number of small holes punched in the metal cap, and when the jar is filled it is given to the colony in an extra hive body that is placed on top of the brood body for that purpose.

First saw a piece of board just the size of the outside dimensions of the hive body, and in the center of this board cut a hole big enough to insert the cap of the jar, the jar having first been filled and cap and rubber band screwed securely on, and lifting the hive lid from its place, put the piece of board referred to over the brood nest with the filled jar in the hole. Then place over this an extra body, and inside of this body pack some old clothes or bags to conserve the warmth of the colony,—and the trick is done; the bees will soon carry the feed below. I have used all kinds of patent feeders, but have found nothing to equal this.

In a couple of days the bees will have carried all the syrup below and have stored it in their combs; if they need more, fill the jar for them again (if a two-quart jar is used, more syrup can be given them at a time).

It is the practise of some beekeepers to resort to stimulative feeding in the spring; that is, to give the colony a little syrup every day for a number of days preceding the honey flow, as it tends to make the bees rear brood earlier and in larger numbers than they otherwise would, and thus have the colonies strong for the honey flow. For this purpose nothing is better than the feeder already referred to.

If the bees have been wintered on their stands outdoors with proper protective packing, it will be best not to remove the packing until settled warm weather makes it safe.

If the bees have been wintered indoors in a cellar, they should not be brought out until winter weather is entirely over. Then when they are brought out, every hive should be given some added protection for some time or until settled warm weather, as this will overcome the possibility of the colony being chilled by the change of atmosphere from the cellar to the outdoor air.

Many beekeepers when placing their bees outdoors immediately wrap the hive with some kind of waterproof building-paper, being careful not to close the hive entrance, and in almost every case it proves effective and prevents spring dwindling, —the bane of the old-time beekeeper.

Be sure to remember the foregoing point, for it is important and one of the fundamentals, and its neglect will surely lead to failure.

The first work among the bees in spring should be short and simply for the purpose of finding out the condition of the colony—whether dead, queenless or short of stores—and should not be needlessly prolonged. As the weather gets warmer and more settled, other necessary work among them can be done.

TRANSFERRING.—Early spring is the best time to transfer to modern hives, bees that have been bought in old box hives, for at that season the brood nest is not large and little honey is present, so that it is an easy matter to drive the bees from their combs; and by cutting their combs to the proper shape they can be fitted in frames and held in place with rubber bands until the bees can fasten them securely, which they will do in a couple of days.

There are various ways of transferring bees from their old box hives to those with movable frames.

Another good way to transfer is to take the old box hive just about swarming time and bore two or three holes in its top and place over it a modern hive with frames of full foundation in it; in a few days the bees will go up into it, and as soon as the queen gets to laying well the bees will carry up the honey from the old hive and when the brood is hatched the old hive can be taken away and its old combs melted into wax. Be sure, however, to stop up every crack around where the new hive fits over the old one. The transfer is accomplished without mussing with honey, brood, etc.

When I say to have the new frames filled with sheets of foundation, I mean the sheets of beeswax that are sold by supply houses, which sheets have been run through rollers and have had stamped

on them the beginning of worker cells.   There
are other methods of transferring, but the two
given are the best, especially the latter method.

INCREASE, HOW TO MAKE, IN THE SPRING.—There
are several ways to increase the number of colonies
in the spring, but as this book aims to give only
the cream I will describe the best two methods in
vogue and give what experience has demonstrated
to be the best.

One method of increase is accomplished by
simply dividing the colony, which enables one to
double the number of colonies in the apiary; if
carefully done this may in many instances also result in a crop of honey as well, but if rapid increase is carried on no surplus will be produced,
as the colonies divided are not sufficiently strong
to build up in time for the honey flow,—but what
is lost that year in honey is many times made up
in added colonies.   You can not make bees and
honey at the same time, if the increase is very
large.

To increase by simple division, take a strong
colony of bees and lift from a ten-frame colony
six or even seven of its frames of sealed-up brood
and queen and place in a new hive and carry to a
new stand; put in front of the entrance a slanting
board so that the bees will mark their new location.

The reason I give the queen and most of the
bees and sealed brood to the division that is moved,
is because many of the bees will return to the old
stand and thus make the division about equal,
while the possession of the queen by the colony
that is moved will keep the bees from deserting it
completely.

The original hive containing the three frames of

bees and unsealed brood is left on its stand, and, from the frames of brood, will rear another queen, —provided there are eggs in the frames, which should always be the case.

Then place in the hive that was moved, also in the one that was left on its original stand, frames of full foundation to fill them out, and the thing is done. If the division is made in the early part of May, both colonies will, if the season is good, produce a surplus and go into winter quarters with good strong stocks.

Whether the increase be made by one division, or many (as will be explained), great care should be used and no increase started much later than June, for if it is, there is danger that the colonies will not build up in time for winter, which is specially the case if the increase is rapid.

I have known of instances where one division of the colony has been made as late as August and both colonies did well, but there was a good big late flow from buckwheat and fall bloom, which made it possible; for the average beekeeper I say, make your increase early.

When it comes to the matter of rapid increase I should advise the beginner to go carefully, for unless it is backed by experience it will result in failure; but in the hands of one who understands the bees it is perfectly safe.

The following is the method that I have practised for years, and though as many as four hundred nuclei, or small colonies, have been started, it has proved to be successful in almost every case.

Sometimes it will happen that after a large number of these nuclei have gotten under way, the honey flow will cease; in which case it will be necessary to feed them for some time, even a month, giving, say, one-half pint of syrup each day, or as much as they will take up over night. For this purpose there is no better feeder than the one before. referred to.

I might add that the supply dealer can supply you with the Mason jar caps with perforations that have been stamped in them by dies, and the jars themselves can be taken from the pantry or purchased at any store.

Select a strong ten or eight-frame colony, and toward evening (say 4 or 5 P. M.) open it, lift out two frames of bees and brood, and place them in an empty hive; then fill the hive out with new frames of comb, or if combs are not on hand, with frames of full foundation. Take two other frames and place in another hive, and so on until from the ten-frame colony you have made five small ones, being sure to see that each colony has if possible some young brood from which to rear a queen. The old queen will of course be in one of the divisions and that will not need any other.

Take each hive so started and move it to the stand it is to occupy, and completely close the entrance with a strip of wood, with a half-inch hole bored in, which close with a cork; the bees are not strong enough in numbers to suffocate. Leave them imprisoned for three days; at the end of which time, draw the cork from the entrance block, which will allow one bee to pass out at a time (and don't forget a slanting board at the entrance).

Watch them carefully, and be sure that they start a queen cell, and if they don't, give them one that has been built in a strong colony whose queen has been caged to make them start, in which case you will get a larger and better queen, and when mated she will begin to lay.

A better way, however, is to have on hand at the time of division a number of queens that have been reared or purchased for the purpose, and when the division is made fill the smoker with tobacco and send a few good puffs down over each little colony and then puff some on the queen to be introduced; drop her down between the frames and close the hive, and in ninety-nine cases out of a hundred she will be accepted at once. The tobacco smoke gives both the bees and the queen the same odor, so that there is nothing about her to indicate that she does not belong to them, and it prevents the possibility of her being recognized as a strange queen and being balled and killed.

Some beekeepers practise feeding half a teacupful of syrup to these small nuclei, right from the beginning. This is not really necessary if there is a steady flow from the fields, but upon its first sign of cessation begin feeding at once; and to avoid robbing among the bees, give the feed late in the afternoon or early evening, when the bees are not flying.

After each increase has marked its new location, the slanting boards can be removed.

Every few days look at the nuclei, and if increasing in strength, take one of the frames of foundation and slip it between the two frames of comb and brood, being careful not to repeat the operation faster than the bees can draw it out and cover it, and in this way every frame will soon be filled with nice worker combs; by fall each of the small nuclei will have built up to a strong colony.

A FEARLESS BEEKEEPER AFTER SCOOPING UP A SWARM
WITH BARE HANDS

Chapter VI

SWARMS AND SWARMING

A swarm in the hive is worth two on the tree.—Vincent.

A great deal has been written on this phase of beekeeping, much that is both nonsensical and confusing; and it is not my purpose to add to the confusion, but rather to give a few simple instructions as to detecting signs of swarming and methods for controlling it.

It must be remembered that swarming is a perfectly natural act on the part of a colony, and is brought about by conditions that prevail at certain seasons of the year. May and June are the principal months during which swarms emerge, though they often come out at other times.

He who depends upon natural swarming for his increase will make a mistake, for in the majority of cases natural swarming will about make up for winter losses, and no more.

During the early part of May much honey is being gathered by the bees and the brood nest is consequently crowded for room. It is this condition that induces the old queen and a large part of the colony to swarm out in quest of more commodious quarters.

Sometimes the colony will give notice of its intention to swarm, by the bees clustering on the front of the hive, though this is not an infallible sign, as bees will do this at other times when the swarming impulse is farthest from them.

The surest way of determining the matter is

by an examination of the brood nest, which will usually be found to be in a crowded condition, the cells filled with honey, pollen and brood, and queen cells started on a number of frames. When these signs are present, prepare for a swarm almost any day, usually between the hours of 9 A. M. to 3 P. M.

When once the swarming fever has reached its highest point, and thoroughly infected the mass, they come tumbling out of the hive like mad and whirl around in the air, their buzzing being heard at even a considerable distance. This they will continue to do for a few minutes, when, joined by their queen, they will generally settle on the branch of a tree or bush, awaiting the return of their scouts who have been looking for

THESE BEES ARE GETTING READY TO SWARM

a home in the heart of some tree or under the clapboards of a house, to which on their return the scouts will escort the swarm. For this reason the swarm had better be hived at once, and the method is both simple and easy.

Sometimes the swarm will cluster in the top of a tall tree or other inaccessible place, and the beekeeper will have to use his wits to meet the condition that will arise. Sometimes a ladder can be used to reach them, and by sawing off the branch or part of it on which they have clustered they can be brought to the ground and shaken in front of an empty hive that has frames with either starters or full sheets of foundation in them. Usually they will enter the hive, and all trouble is over.

When swarming, bees can generally be handled without either veil or gloves, as they seldom sting at this time; due to the fact that they filled up with honey before swarming, which has so distended their abdomens that they can not contract themselves to sting.

With the swarm has gone the working force of the colony, and for this reason it is best to take all surplus tops in which they had been working or storing honey and place same on the hive which the swarm has entered; experience teaches that the hive out of which the swarm emerged will seldom produce any more surplus, for by the time the brood therein has hatched and become field bees, the honey flow will be about over.

Sometimes the hive that cast the swarm will, upon the hatching of the queen cells that were left in it, cast another swarm known as an "after swarm," but as these are small and undesirable it is best to prevent them by cutting out all queen cells but the best one in each hive, thus thwarting this possibility. Be sure, however, in a few days to see that this cell has hatched and that the queen is laying, for if for any reason it should fail to hatch, the colony is without the means to rear one, as the old queen has been gone so many days that there are no eggs sufficiently young for this purpose.

Some beekeepers early in the season see that all queens in the yard have their wings clipped, and so when the swarm comes out the queen, unable to join them in flight, is found hopping about in the grass in front of the hive and can be caged until wanted later.

No attention then need be paid to the swarm, for though they may go a mile from home and cluster on a tree, and a neighbor shake them into a hive, they are sure to return as soon as they discover that their queen is not with them, and for this return you should make ready.

A GOOD "CATCH"; DON'T DELAY
IN GETTING THE SWARM
INTO A HIVE

After the swarm has gone, take the hive out of which it came and carry it to some other location if increase is desired, and in its place put on the old stand an empty hive with frames, foundation, etc., and when the swarm returns it will immediately seek the old location and begin to enter the hive, the bees making a sound with their wings that is pleasant to hear. When about half of the bees have entered, the caged queen can be gently dropped upon the entrance board and she will quickly enter; and thus the swarm "hives itself," without a lot of worry and climbing on the part of its owner.

The clipped queen can be picked up with naked fingers, she will not sting; but be careful not to injure her, as she is very tender and delicate at this time, being filled with eggs.

A pair of curved manicure-scissors are about the best thing with which to clip a queen's wings. It can be done at one clip, as she moves slowly over the combs, without even touching her with the hands or without fear of cutting off a leg as often happens when ordinary scissors are used.

SWARM CONTROL, OR HOW TO PREVENT SWARMING. —I confess that I feel like the majority of beekeepers when I say that it causes me no pleasure to have my bees swarm, and for this reason many methods and devices have been employed by me to prevent it.

The beekeeper who has several out apiaries is often worried with the thought that some of his bees remote from home may swarm and the swarm get away, as it hardly pays to keep a man on watch all the time; so there has been much discussion at the different bee conventions as to how to control swarming, for in rare instances only has it been entirely prevented. Sometimes in spite of all we can do a swarm will occasionally come out, but we are able to keep it in bounds so that it ceases to be a nuisance.

Some of the things used to control swarming have to do with the hive used, while others have to do with its manipulation.

Personally I have no faith in patent swarm self-hiving hives, and believe it to be money wasted to purchase them; nor have I any patience with those who offer to sell secrets for $1 about " How to

Attract Swarms," for in most instances the so-called secret simply says to place a lot of empty hives in the woods, first rubbing their inner sides with anise-seed oil and tacking a little red flag to each hive to notify passing swarms that here is a home.   Experience has shown that a dollar given up for this or any other so-called bee secret, but illustrates the truth of the adage which says, "The fool and his money are soon parted."

THIS SWARM
TOOK POSSESSION
OF A
BIRD HOUSE

Swarming can largely be controlled by using large hives, giving them a wide entrance for ventilation, and by giving added storage room for the surplus honey as it is gathered; this is the " secret " if it can be called such.   In addition to this it is a good practise to cut out all queen cells that have been started, not overlooking even one, and giving more room to the colony.

The foregoing method has proved very successful where extracted honey is produced, for the constant extracting of the honey from the combs as soon as they are filled and sealed, keeps the colony from being crowded for room.

When it comes to the production of comb-honey it is not so easy a matter, for the combs are not extracted but must be left on the hive till finished, and this fact causes bees kept for comb-honey to swarm more than those kept for extracted honey.   Adding storage boxes, cutting out queen cells, ventilation, and shade, will in a measure tend to keep swarming within bounds when producing comb-honey, but even then it will constantly be a menace, especially in the out apiaries, and this is why most out apiaries are run on the extracted plan.

There is, however, a method for controlling swarming even when working for comb-honey, and it is what is known as the " Shook" or "shaken swarm" plan, and simply swarms the colony artificially, at the beekeeper's convenience, and not when it suits the bees. The plan is as follows:

A few days before a colony swarms, having first shown signs of an intention so to do, remove it to one side of its stand, and in place of it put on the old stand a hive exactly like it, filled with frames with nothing but narrow starters of foundation in them, under no circumstances using full foundation or combs. The bees from the old colony are then shaken from their combs on to the entrance of the entrance board of the new hive; and if the weather is warm, almost all if not all of the bees can be shaken or brushed from the combs of the old hive.

The surplus bodies of the old hive are placed on top of the new hive, being sure to place a queen excluding board between the hive body and the surplus bodies to confine the queen to the starters below. Having no combs in the brood body in which to store their honey, the bees will store it in the supers with a vim, just where the beekeeper wants it.

In some localities where the flow is short this plan will insure a good crop where otherwise such a surplus would be impossible. If increase is wanted, the old colony can be removed to a new location and a queen given to it. If no increase is desired, the old hive can be left standing beside the new hive, and, as the brood hatches, the bees

from it can every few days be shaken in front of the
new hive, being careful to cut out all queen cells
from the old hive for the first ten days.

When all the bees have been shaken from the
old combs after they have all hatched, the combs
can be cut from the frames and melted into bees-

wax and a hand-
some return re-
ceived from them;
but when cutting
them out, be sure
to leave about an
inch of comb at-
tached to the top
bar of each frame,
as it will serve for
starters the next
season when the

SMOKING A NEWLY HIVED SWARM TO
HURRY THE BEES IN

same frames are used for shaking bees on to them
in another hive.

For comb-honey the foregoing plan is without a
peer. Those who advocate it say that it prevents
swarming because the shaking of the bees in front
of the new hive convinces the bees that they have
swarmed, and thus cures the fever. This is wrong;
the reason they do not swarm is simply because
a colony will seldom swarm from a hive until that
hive is filled with combs in the brood nest, and by
the time they have drawn the starters out to full
combs the flow is usually over and the swarming,
too.

Some beekeepers practise dequeening their
colonies just before swarming, and by cutting out
all queen cells a few days before, prevent it.
While the absence of the queen under this method

means a cessation of brood rearing during this period, yet its advocates claim that there is no real loss, as the bees that would be reared during this time would come on after the flow is over and be consumers. Although this is true, yet I know that a colony will work with more energy with a laying queen than without one.

Mr. L. A. Aspinwall, of Jackson, Mich., has in a quiet way for the past twelve years been carrying on some experiments with a hive of peculiar construction for comb-honey, and during all that period has never had a swarm, and has produced large surplus through ability to keep the entire working force in the hive.

These hives are not on the market yet and may not be for some years, but my opinion is that they will eventually be universally adopted. The frames are closed-end ones, and where the comb ends the frame for a distance on each end of four inches is slatted, which permits a clustering space for the colony when crowded, and yet the slats prevent the bees from building burr or brace combs, which they certainly would build were the space left open.

Just before the swarming season, these brood frames are spread apart and between each of them is slipped a slatted dummy frame of the same size, which gives additional clustering space in the hive, and as the frame so inserted has no comb in it, but is filled with slats just a bee space apart, it enables the beekeeper to give the colony all the room it needs and removes the crowding that produces the swarming.

CASE FOR CARRYING HONEYCOMB TO EXTRACTING ROOM

Chapter VII.

COMB AND EXTRACTED HONEY

A little honey now and then is relished by the best of men.
—Revised Proverb.

Having satisfied his bee fever by the possession of some bees, and having safely brought them through the winter and by careful management made the colony strong for the honey flow, the beginner is ready to make preparations for a crop of honey.

If the object is to produce comb-honey, the super chambers with sections must be ready to place upon the top of the colony as soon as the bees have begun to store new honey in the tops of the brood

A NETTING TENT WILL KEEP OFF ROBBER BEES WHEN NO HONEY IS COMING IN AND HIVES MUST BE OPENED

frames, which can be easily determined by seeing when the combs are being capped over with new white wax. Generally speaking, about the early part of May, or during apple bloom, is the proper time to place the upper stories on, one at a time.

Under all circumstances use full sheets of super

foundation in the little section boxes, as the bees will enter them much more readily than if only starters are used. Fasten each sheet in securely by aid of one of the fasteners sold for that purpose, leaving just a little space at the bottom of the section box to allow for sagging.

When the super is filled with sections properly

HAND EXTRACTOR IN OPERATION

prepared, lift the lid of the hive and place the super on it, putting the hive lid on the super,—and all is ready.

In a few days carefully examine it and if the honey continues to come in and the section boxes are nearly filled, prepare another super for the hive in the same way and add it to the hive, placing it between the super nearly completed and the hive body proper, but never put it on top of the first super.

If the flow continues, it is possible to place as many as four supers at a time on a strong colony, adding them only as needed.

As soon as the combs are capped over remove them at once; nothing is gained by leaving them on till the end of the season, as they become travel-stained and the white caps unsightly.

The best way to take the supers off is to place a board with a Porter bee-escape in it between the hive body and the supers, and in twenty-four hours the bees will have gone through it below and the supers can then be taken off without the bees to bother.

Some beekeepers simply blow clouds of smoke down through the supers and in this way drive the bees below, but it is a poor practise and is open to the objection that it makes the bees uncap some of the cells to fill up with honey.

When removed, the sections should be scraped of any propolis or bee-glue that has been deposited on the little boxes and the honey placed in a warm, dry place where the bees can not get at it. Under no circumstances store it in the cellar or other cool place, for this will make the surfaces of the combs sweat and drip, but store it in the kitchen pantry or in the attic, as these are admirable places.

Much could be written on this phase of honey production, but it would only tend to confuse; and experience proves the foregoing method to be the best.

EXTRACTED HONEY.—In the production of extracted honey the method is in some respects similar to that used in comb production, only instead of using a super containing small section boxes, full hive bodies are used.

By all means use a ten-frame hive in producing extracted honey, as it gives larger storage space and the hives do not have to be tiered up so high.

As soon as the same conditions prevail in the brood nest as in the case of preparing for comb-honey, place over the hive body an extracting super, or full-sized hive body fitted with frames of comb; or if these are not on hand, frames with full sheets of foundation (being sure that the combs or foundation are wired in to prevent their being torn from the frames in the extractor).

Between the hive body and the extracting super

be sure to place a queen excluding board (which is a piece of zinc with perforations in it sufficiently large to permit the workers to pass through up into the super, but too small to allow the queen to pass through), for should she get above and rear

HANDING COMBS OF HONEY INTO THE EXTRACTING HOUSE

brood in the extracting super it will be in the way when in the extractor, and liable to be thrown out. (NOTE: A queen excluding board should also be used when producing comb-honey.)

Perhaps the majority of beekeepers use but one extracting super on a hive at a time, extracting as often as filled, but it will be well to have an extra one handy for each hive in case of a heavy flow, or to leave on for the thorough ripening of the honey.

The time to extract is when about two-thirds of

the cells in the frames are capped over, although some beekeepers leave the supers on till the end of the season, claiming that the honey ripens better, but I have never seen any advantage in so doing; most bee men extract as soon as ready, usually twice or oftener as occasion requires.

A good practise is to extract during the early and the latter part of July when the clover flow is over and then again in the fall after the fall flow, for in so doing the clover honey, which is lighter in color and of a more delicate flavor than the fall honey, is kept separate and is sold for a higher price. (This applies also to comb-honey.)

VATS FOR HOLDING HONEY TILL PUT IN BARRELS FOR SHIPMENT

When the combs are ready to be extracted, take an empty hive, place it on a wheelbarrow and cover it over with a wet towel (which will keep robber bees out) and go to the first hive, smoke the bees a little, and then lift the combs out one by one shaking all adhering bees on the entrance board of their hive, using a bee brush to clean them off entirely. These combs are then placed in the empty hive and covered, and carried into a room where the extractor is to be worked; with an extracting knife the cell-caps are shaved off and the frames placed in the extractor and whirled around until all the honey is extracted, when they are returned to their hive again, either immediately or at the close of the day when bees are not flying.

Extractors are two, four or eight-frame size, but unless one has an apiary of fifty or more colonies a two-frame extractor will suffice, although there are advantages in having a four-frame one, as it extracts more frames for the same amount of work.

POWER EXTRACTOR

A piece of cheese-cloth may be tied over the honey gate at the bottom of the extractor and the honey can be run into kegs, cans or buckets as extracted, and sealed up to keep it away from the bees. Great care should be exercised, however, to see that the extracting room is bee tight, or a good case of robbing may be developed.

Many beekeepers store their extracted honey in sixty-pound cans, two of which come in a case; and although this is a very satisfactory way of storing and shipping extracted honey, these cans are quite expensive; they cost eighty cents for a case of two.

Other honey producers buy from the manufacturers half-barrel kegs holding one hundred and sixty pounds of honey. These kegs can be bought, delivered to the beekeeper, for about forty cents each, and being new the honey is not contaminated. The various styles of honey packages will be further discussed under the chapter on Marketing Honey.

It is a question as to how many frames of comb should be in the extracting super, but it is a good plan not to have the super crowded with as many as it would take if used for a brood body.

If the super is a ten-frame one it will be a good thing to put but eight frames in it and spread them well apart, so that the bees will have plenty of room to cluster and store honey; by so doing the combs will be fat and, when uncapped, provide a good surplus of beeswax from the caps and combs.

MATING BOXES USED IN QUEEN REARING

Chapter VIII

QUEEN REARING AND INTRODUCING

" A bee has the right to have the best of ancestors ; blood will tell."

While it is true that the majority of beekeepers prefer to buy from queen breeders what queens they need, yet a great many rear their own queens, especially if they have a great honey - producing strain of bees which they desire to perpetuate.

If left to themselves a colony would rear queens only at the swarming season, ARTIFICIAL CELL CUPS READY TO RECEIVE GRAFTED EGGS or when made queenless, or when about to supersede an old queen that shows signs of failing.

When a colony casts a swarm it will generally be found that there are a number of nice cells left in the hive from which they swarmed, and as these cells are built under the most favorable conditions they are generally the best. To let the colony tear down the surplus ones when the first virgin queen hatches would be the height of folly, as a little care will result in securing some of the very best queens imaginable.

To do this, cut out every cell but one, being

careful not to open or crush the cell in so doing, then place each cell in a West cell protector and

CELLS COMPLETED BY QUEENLESS COLONY

stick the protector on the comb of some strong colony, and when it hatches in a few days the queen can be mated from a small nucleus or else given as a virgin to increase, or to a queenless colony and allowed to mate from her permanent home. In this way a nice lot of queens can be reared each season to supply the needs of the yard.

If a considerable number of queens are needed, it will be necessary to use some inexpensive appliances for the purpose, and then they can be reared in unlimited numbers.

First select a strong colony and take from it their queen, and in about two days it will be ready to begin cell building. Then from that colony, or some other, select a frame of brood not more than three days old, and with a small grafting needle gently lift one of the larvæ from its

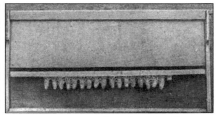

FRAME OF COMPLETED CELLS

cell and place it in the little artificial cup made for the purpose.

After a number of these have been grafted, place them in the frames made for such purpose, and deposit the frame of cell cups in the center of the queenless colony. In about fourteen days take the cells from the frames, put each one in a little cage by itself and fill a frame full of these cages and

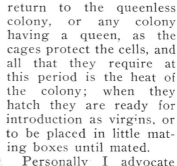

return to the queenless colony, or any colony having a queen, as the cages protect the cells, and all that they require at this period is the heat of the colony; when they hatch they are ready for introduction as virgins, or to be placed in little mating boxes until mated.

Personally I advocate giving to such colonies as need them, virgin queens, as they are more read-

CELLS IN PROCESS OF MAKING

ily accepted, and I thus avoid the trouble of looking after a lot of little mating boxes.

Mr. E. L. Pratt, of Swarthmore, Pa., is the originator of the Swarthmore queen-rearing system, which has proved a great success in the hands of those who have tried it, its simplicity rendering it effective even in the hands of a beginner.

If from any cause a colony has become queenless the condition can be remedied by giving the colony a frame of young eggs and permitting them to rear their own queen. As this involves a wait and loss of brood, a good many prefer to rear or buy queens and give them one at once.

If the queens are bought, the directions for introducing are printed on the mailing box in which they arrived; but for the benefit of those who pre-

CELL PROTECTORS IN PLACE

fer raising them, I will describe two methods of introduction that have proved to be the best.

Whether the queen has been reared or bought, she should be taken in the little introducing or mailing cage and (with honey candy sufficient to keep her, stored in the box) placed between two frames of brood or bees in the colony to which she is to be given. Of course it is understood that the colony to which she is to be introduced is at the time queenless, and has been so for at least three or four days or even longer.

This honey candy is made by mixing to a stiff dough some extracted honey and pulverized sugar. The candy fills up the space between the queen and the little hole in the cage through which she is to escape, and by the time the bees have eaten away the candy so that she can get out, she will have acquired the scent of the colony and will be accepted, as the odor of their hive is the only means by which the bees know their queen. In a few days she will be found walking over the combs, laying, when the cage can be taken out for good.

CUTTING OFF FINISHED CELLS

Another method by which the queen can be immediately introduced, is to fill the smoker with strong tobacco and give the queenless colony a good smoking every five minutes till it is well scented, being careful not to give it too much (as too much is worse than too little), and then puff a few whiffs on the queen in the cage, and gently drop her on top of the frames,—and in a jiffy it is done.

I use this method exclusively and seldom lose a queen, with the advantage of having her introduced at once; in fact, I have taken a queen from a colony and in five minutes introduced another in this way.

Queenlessness is too serious a loss to a colony to be allowed to continue, and upon its discovery should be remedied at once. It may be discovered in several ways.

One way is to examine the combs, and if there is no brood or eggs present it is reasonable to suppose that the queen is gone, although sometimes you will be fooled,—as in the case of a hive having a virgin that has not yet begun to lay.

Not finding the brood, then look over the combs for her majesty, and if careful search fails to reveal her, it is pretty certain that the colony is queenless.

Unless young eggs are given to them or a new queen provided, there is danger that they will introduce a laying worker, with disastrous consequences to themselves. They will select one of the workers and feed her stimulative food, and she will lay, but all of her eggs will prove to be drones; and the eggs will be deposited irregularly over the comb, sometimes as many as three in a cell.

In most cases, giving the colony some young brood will make them raise a queen, and the laying worker will quit. But sometimes they will refuse to rear a queen from the brood given, and will even kill a queen given them; in which case it will be well to move the hive now depleted in numbers, and on its stand place a strong colony; taking the queenless colony a few yards away and shaking the bees off in the air, when they will return to the colony on their stand and, if the honey flow is on, will be allowed to enter and go to work. The lay-

SECURING ATTENDANTS TO CARE FOR QUEEN DURING
HER JOURNEY BY MAIL

ing worker, if she enters that hive or any other, will be put to death by the workers.

It would be impossible in this brief treatise to go extensively into the question of queen rearing, as there are a number of books on the subject; but I have outlined enough to enable the average bee-keeper to go ahead and rear the best of queens.

QUEEN REARING. CELLS AND CAGES ON TOP OF OBSERVATION HIVE

CHAPTER IX.

OUT APIARIES. MOVING BEES

Don't bite off more than you can chew.—Tim.

Unless the beekeeper is located in an unusually favorable locality, he will have to resort to a system of out apiaries, especially if he depends upon bee-keeping for a livelihood.

Generally speaking, it is the height of folly to overstock a locality with too many colonies, for there is a limit to even the most promising districts. It is the lack of bee forage that has led the large beekeepers to adopt the plan of having several yards, ranging about three miles from each other, and thus securing a splendid return from each yard; whereas, if all were located in the home yard, there would be little or no surplus, and in the majority of cases the beekeeper would have to feed for winter.

As a rule bees will not go more than three miles from their homes for honey, which gives them a circle of territory about six miles in diameter; so for this reason it will be best to have the yards no nearer to each other than three miles.

For instance, one yard could be located at the home, and a yard each three miles in opposite directions from the home yard, so that no out yard would be more than three miles from home, and all within easy driving distance.

A small piece of ground can be bought or rented for the out yards and a small building erected for a work house in which to store hives and do the extracting.

Even though a building is not erected, it is a wise thing to have at least some kind of a box or storage place there in which to keep smoker, veils and tools, as each yard should have its own working

A HIVE ON SCALES HELPS TO KEEP TAB ON THE APIARY

tools, independently of each other, and thus avoid all possible delays in getting to work when once the out yard is reached. The tools for the out yards should be of the same character as those used in the home yard, and a good supply of smoker fuel should always be on hand.

He who begins to establish out yards should keep constant account as to just how much each yard is gathering, for while one yard may be storing a goodly surplus, another may need to be fed; for this reason it is a good plan to have in each yard a hive upon scales, so that as its weight increases it is a fair indication that the colonies in this particular yard need more room.

If extracting is practised as soon as the honey is sealed, it will be necessary to have some room or wired-in place where the extracting can be carried on in security from robber bees.

The out yards require constant watching, for should the bees become crowded in their hives, and the beekeeper fail either to extract or to give them added supers, a lot of valuable swarms would get away in the beekeeper's absence, which would be a distinct loss.

If the bees are wintered outdoors with proper protective packing, a lot of work will be saved each year, as it would be quite a job hauling the bees back and forth to the home for cellar wintering. For this reason I have adopted exclusively the outdoor wintering of bees, both at my home and at out yards; and as the packing is ample and effective it also prevents the spring dwindling so common when bees are taken out of their winter cellar in the spring. (NOTE: The best way to pack bees in either the home or out yards will be fully treated under the chapter on How to Winter Bees.)

The question as to the number of out yards a man can profitably run, will depend entirely upon the man's ability as a manager and his adaptation to the work.

There are some people who can successfully manage a few colonies at the home yard, who, if they should launch out in a system of out apiaries, would make of it a monumental failure; so for this reason my advice would be to " go slow," work into it gradually and let your growth be in proportion to your success.

There are some beekeepers who each year move their apiary from place to place, following the

A WELL-KEPT OUT APIARY

bloom; but as this method is precarious at best, I should not advise its adoption except in rare instances, and even then only in the hands of an expert.

In moving bees to and from the out yards, a great deal of caution should be exercised; for should the hive be rudely jarred and the bees escape while in transit, disastrous consequences will surely follow, and possibly the horses may be stung to death.

The hives may be prepared by having wire

screen tacked over their tops and bottoms the
night before, and then in the cool of the morning the
hives should be placed on a wagon having springs
and the body filled with hay or other soft and
yielding material. Personally I use a frame of
wood seven-eighths of an inch thick and on one
side of this frame I tack wire netting like a window
screen; a few nails will hold one in place on the
top and the bottom of the hive body. They can
be used indefinitely and are also just the thing for
shipping bees away by express or freight.

Above all things, have the hives so prepared and
arranged on the wagon that there shall be ample
ventilation while in transit, and when the out yard
is reached unhitch the horses and take them some
distance away until every hive is safely lifted from
the wagon.

It is best not to liberate the bees till toward
evening. In case no bees have gotten out the
horses can again be harnessed to the wagon and
the wagon drawn away some distance, till the
screens can be taken off, the bees liberated and
each hive left in position with bottom board and
lid in place.

EXTRACTING A BEE STING — EASILY DONE

Chapter X.

BEE STINGS. REMEDIES

An ounce of prevention is worth a pound of cure.—Martha

From a mistaken notion that bees are naturally vindictive and sting without the slightest provocation, many people are deterred from having them on the place at any price.

As a matter of fact, a large number of stings that annually come to beekeepers as the tribute paid for the profits of the work, might with a little care be avoided. A lot of irritating causes can easily be removed, and preventives be employed to reduce stinging to a minimum. For instance, bees are not so cross when I don a white cotton suit instead of dark woolen clothes.

Another cause of frequent stings arises from the fact that the beekeeper often goes right from the barn to the bee yard; and the odor of cattle, especially that of the horse, is sure to make bees mad, so for this reason I make it a point to wash up and don my white painter's suit.

Many times I am stung because I am in a hurry to open a hive and do not smoke it properly and wait a few minutes before opening it; a little patience would have prevented a sting.

Kicking or bumping against hives will also make bees angry; a little care in this respect would remove to a large extent the stings we all receive.

The proper way to open a hive is to blow in a few puffs of smoke at its entrance and then pound upon the cover a few times, which so alarms the bees that they fill up with honey immediately; this so distends their abdomens that they are physically incapacitated from stinging. The next move is to pry open the lid just a little bit and puff some smoke over the frames, placing the lid on again, and in a few minutes the hive can be opened and the bees will be as quiet as quiet can be.

It's the quick "jerk and jump" methods of opening hives that account for so many stings, and not because the little fellows are malicious.

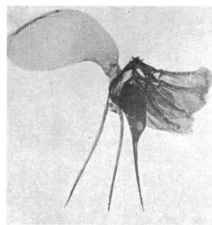

THE STING (GREATLY MAGNIFIED IN THIS PICTURE) IS FEARFULLY AND WONDERFULLY MADE

A good plan to save time is to take a row of hives that are to be examined, and beginning at one end go down the line smoking ten or fifteen hives in succession; and by the time the last one has been smoked the first one will be ready to be opened, and all the rest in their order.

I always use a veil but work with bare hands, as I don't mind an occasional sting on them; whereas one on the face might be very painful and might close the eyes and compel me to stop work for a while.

If, however, you do get stung, don't be foolish enough to try to pull the sting out with the fingers, for in doing so you will be certain to force more poison into the wound. The proper way is to take a knife blade and push it out; and if you will then take the smoker and puff a lot of hot smoke on the place, until it gets so hot as to be almost unbearable, you will find that the hot smoke will act as a counter-irritant and in most cases take away the pain and prevent swelling. This is a little trick I learned by experience, so paste it in your hat.

If we must get stung we may all take satisfaction in the knowledge that the poison of the bee sting is an invaluable remedy for rheumatism, as the formic acid of the bee off-sets the uric acid of the disease.

It has been known for many centur-ies that, general-ly speaking, bee-keepers are im-mune from this affliction, due, no

STUNG! HE'LL BE CAREFUL NOW

doubt, to the agency of the bee sting; and there are a number of reputable preparations on the market made from the stings of the honey-bees, for the cure of rheumatism.

While there have been rare instances where the sting of a single bee has produced death, yet in the majority of cases the pain is but for a few minutes and the inconvenience slight. In any case, an ounce of prevention is worth a pound of cure.

As far as remedies are concerned I have but little faith in them. I have never known an instance where the application of mud to the part stung has made any great difference. As before stated, the puffing of hot smoke on the place, after the sting has been scraped out, seems to be about the best remedy.

I know a beekeeper who declares that frequent applications of a solution made from equal parts of tincture aconite, chloroform and laudanum have proved very helpful; no harm can be done by giving this a trial.

If the part stung should swell a great deal and become painful and feverish, the application of an ice bag, or cloths wet with cold water, will feel very comfortable and reduce the swelling and remove the burning.

If the sting is on the hand much relief will be experienced by holding the hand in very cold water from time to time.

It is a fact that almost anyone may, from being constantly stung, become immune to the effects of the bee sting.

Be quiet in all your movements about the bees; never dodge or strike at a bee whose constant buzzing before your face may but be an expression of curiosity on its part; a sting is almost sure to follow a quick move on your part. If a bee viciously follows you about and shows signs that he expects to sting, just take a shingle and hit him and kill him; but be sure of your aim.

Chapter XI

HOW TO WINTER BEES

Be sure to tuck the little folks in a warm bed for their long nap.—Harriet.

There are two plans in vogue among northern beekeepers as to wintering bees, some practising the outdoor wintering, while others store their bees in the cellar; and as each plan possesses real merit I shall describe them both.

The arguments advanced in favor of indoor wintering are that it does away with the trouble of having to pack for winter every hive outdoors; also, that it is a distinct saving in honey, as bees wintered indoors need only about half the stores for feed as compared with those left outside; and that the danger of winter losses is greatly reduced.

Personally I have found that the packing of bees for winter is no more troublesome than the work of carting them in and out of the cellar, while the extra amount of honey required for outdoor wintering is more than made up by the absence of dwindling in the spring.

It must be said, however, that the majority of professional beekeepers prefer to winter their bees in a cellar or special repository for them, even though it entails a great deal of work to carry them in and out of the same.

If the colonies are to be wintered indoors, they should be carefully looked over before cold weather sets in. Every colony should have at least fifteen or twenty pounds of good sealed-over honey, and if any are light they should have their needs supplied by taking combs of honey from extra-heavy colonies that can stand it; but if these are not at hand then they should be given a syrup made of equal parts hot water and pure granulated sugar, fed to them over the brood nest from a glass jar with perforated cap as described in another part of this book.

All queenless colonies should be given queens; and if these are not obtainable, the queenless colonies should be united each with a "queen right" colony that may be weak in numbers; and in all cases unite all weak colonies, as a weak colony has little chance of coming through a winter whether wintered in or out-of-doors.

PREPARING HIVES FOR OUTDOOR WINTERING

The time to place the bees in the cellar is after real cold weather sets in, say about the 15th of November for the northern states; but each bee-keeper must be governed by the weather conditions of his locality, the time mentioned being simply given as a general guide.

Almost any good cellar where the temperature ranges between 45° and 65° will be a good place to winter them, and a slight variation in temperature will do

ONE WAY OF PROTECTING HIVES IN WINTER

no material harm. But be sure to tack some tar or building paper over every window to exclude all light, for the bees must be wintered in absolute darkness. It is a good plan to partition off, with building paper or otherwise, a part of the vegetable cellar and put the bees in it.

Toward evening gently carry the bees, one hive at a time, into the cellar; with each hive on its own bottom board, with full entrance open and lid in place; put them in orderly rows and pile them one on the other even though they reach to the top of the cellar. After they have all been placed in the cellar don't tamper with them any more, as it only tends to make them restless and does no good whatever.

WINTER IS HERE, BUT IT HAS NO TERROR FOR A COLONY WELL PROTECTED AND WITH AMPLE FOOD

The only things you have to do now are to keep informed about the temperature, keep them in darkness, and occasionally ventilate the cellar as needed.

The best way to ventilate is to let some cold air come into the part of the cellar that is outside the partitioned-off bee room and then shut the cellar window; when the fresh air admitted has been tempered a little, open the door to the part where the bees are stored and let the tempered fresh air go in to them.

If there is no way to do this as outlined, the

next best way is to wait until after dark and then open the windows of the bee part of the cellar, and for an hour or so allow the outdoor air to come in; then shut all up again.

If, toward the spring of the year before it is safe to put the bees out again, there should come some warm days that make the bees restless to get out, then it may be necessary to open all windows after dark and leave them open all night, making sure to close them again in the morning before daybreak, as the entrance of light will make the bees fly out into the cellar.

HONEY AND PULVERIZED SUGAR MIXED TO A STIFF DOUGH MAKE CANDY FOR BEE FOOD

The time to take the bees out of the cellar will have to be determined by the peculiar climatic conditions surrounding each beekeeper; but as a general guide I should say that some northern beekeepers never take their bees out until the pussy-willow is in bloom, while others go by the calendar and weather, putting them out any time from March 5th to April 10th.

When taken out in the spring the colonies need not necessarily be placed upon the stands they occupied the season before, but can be placed anywhere if they are put out toward evening.

OUTDOOR WINTERING. — The second plan for wintering, and the plan practised almost universally by the small beekeeper in the North, is the outdoor wintering plan. While it is true that a great many people leave their bees out on the stands without any added protection, and in many cases thus winter them successfully even in old box hives, yet there occasionally come winters when they lose all their bees as a result, when a little packing would have entirely prevented the loss.

Perhaps the easiest and least expensive way to prepare the colonies for outdoor wintering, is to wrap each hive in about ten thicknesses of old newspapers, leaving the entrance open, and tying a string about the hive to hold the paper in place. Over this should be placed a box

MOLDING CANDY INTO SHAPE FOR WINTER FOOD

which will telescope, and be water and wind tight and fit snugly over and about the paper. I have practised this plan many winters and have had great success with it.

Another and still better plan, although a little more expensive the first season, is to make a square case of boards that will have a space of four inches between it and the hive body and fill that space with shavings, sawdust or chaff; having the case made so high that there will be at least four inches of packing between the cloth over the frames and the upper edges of the square case; and when the case is full there should be a telescoping cap to go over the top of the case, all of which should be made wind and water tight with some good roofing-paper covering.

Of course you must so make your case that the bees will have their hive entrance open at all times. These cases with their caps can be left on all summer, taking out the packing down to the top of the hive body, and are the best things with which to winter bees outdoors that I know of; when properly made they will last for years.

By this method the bees can fly out and empty their bowels during warm days in winter, and thus avoid bowel trouble so prevalent with colonies wintered indoors; and all the fuss and bother of carting bees is done away with. This is the method I now use exclusively and I find it satisfactory in every way.

Prepare your colonies for outdoor wintering with as much care as outlined for cellar wintering, with the added

SAVING A COLONY BY GIVING IT A CAKE OF HARD CANDY OVER THE FRAMES IN MID-WINTER

precaution of seeing that every colony has at least twenty to twenty-five pounds of stores, as the bees will require more when wintered outdoors than if placed in the cellar. A method of feeding candy to bees when necessary is plainly illustrated in this chapter.

WINTERING IN THE SOUTH.—Colonies wintered outdoors in the northern states require special packing as indicated; in the southern states this added protection is not at all necessary, although it will do no harm.

In southern sections of the country, where bees fly more or less all day, the single-walled hives will answer very well.

The main thing to guard against in the warmer climate is that the bees do not run short of stores, as they will be rearing more or less brood all winter.

In any case see that the lids are water and wind tight, and if possible have the hives sheltered by a windbreak of boards or trees.

Even in the South the bees will be a trifle more comfortable if a super containing a chaff cushion is placed over them in an empty super, although this is not universally practised.

BEES ON BOTH HANDS, BUT WHO'S AFRAID?

Chapter XII

DISEASES OF BEES

The time to cure bees is before they are sick.—Tim.

If care is taken to keep each colony strong, with a good laying queen at its head, the bee-keeper has little to fear from diseases to which bees are prone. It is well, however, for the bee-keeper, whether the owner of few or many colonies, to be familiar with the prevalent bee diseases; and should they appear he should be able to head them off at the very start, or better still, prevent their appearance entirely.

In all the years that I have kept bees, I have never had a diseased colony in any of my apiaries, and this is the experience of the majority of bee men.

There are some diseases like foul and black brood that are very contagious, and when once they have started precautions should be taken to prevent their spread, and this can in the majority of cases be accomplished by energetic measures.

Bees are like human beings, and if kept strong and well nourished by good food they will escape disease entirely.

Bee Paralysis.—This disease is more prevalent in warm climates than in cold ones, although some beekeepers, even in cold climates, have noticed once in a while a colony or two that have been affected by it; but, fortunately, it is a disease that seldom spreads in the North.

The bees affected have a black and greasy appearance, and in the early stages an occasional bee will be seen running around on the alighting board, with its abdomen swollen. Sometimes the bees will walk unsteadily with trembling in the legs, and will of their own accord leave the colony and crawl off in the grass to die.

Re-queening the colony will often result in a cure, although in the South and West this method seems to have but little effect.

The most effective plan is to sprinkle powdered sulphur over the combs of the diseased colony, and should this fail to stop the trouble, then split

the colony up, giving a frame each to a number of strong colonies; and as the sulphur will have done its work, no spread of the disease need be feared.

FOUL BROOD.—This is a disease to be feared when once it has gotten under way,

EIGHT HUNDRED COLONIES ARE WINTERED IN THE CELLAR UNDER THIS HONEY HOUSE, AND THEY KEEP VERY FREE FROM DISEASE

as it will quickly spread from hive to hive, especially if healthy bees from other colonies steal some of the infected honey from the diseased colony.

The presence of this disease is easily detected even by the beginner, as its odor is very offensive. The brood fails to hatch, while here and there are cells of sealed brood, the cappings of which are sunken and of a very dark color.

I would impress it upon the reader that he should not confuse chilled brood with foul brood, for they are not similar in any respect. (NOTE: The reference to chilled brood needs no further explanation than to say that it is not a disease, but a condition brought about by opening the hives when the weather is too cool; it results in loss of brood. Whatever brood is chilled is taken care of by the bees, as they will carry it out of the hives. To avoid chilled brood the beekeeper should be cautious about opening his hives in cold weather.)

To determine whether it is foul brood, take a match and stick it into the cells that are sealed and whose caps are sunken, and if the contents of the cell are ropy and stick to the match it is without doubt the real thing.

While many methods looking to a cure have been tried, the only satisfactory method seems to be to shake all the bees of the colony on to new frames with full sheets of foundation on them, using either a new hive for the purpose, or one that has been thoroughly cleansed by washing with a strong solution of hot carbolic-acid water.

The old combs can then be melted up for wax, and the resultant wax will more than pay for the new foundation.

BLACK BROOD.—This disease, sometimes known as " European foul brood," differs in appearance from the American foul brood already described, and first made its appearance in this country in New York state.

The disease is seldom if ever ropy, has a watery consistency, seems to be confined entirely to the dead grub in its cell, and varies in color from a pale to a very dark brown. In foul brood the larvæ after death dry up and adhere very tightly to the lower side of the cell; but in the case of black brood the dead grub never adheres to the walls of the cell.

To Mr. E. W. Alexander, of Delanson, N. Y., belongs the credit of discovering a cure for the disease which many years ago devastated thousands of colonies in New York state. Each affected colony is made queenless and left so for three weeks, until all the brood has hatched; and in anticipation of a new queen the bees of the colony will clean out and polish every cell for the new queen to lay in, and thus the disease is eliminated entirely,—as I can certify after a careful examination of the colonies so treated.

In treating diseased colonies do all the work late in the evening when no bees are flying, for should there be any robbing, the entire apiary is in danger of becoming infected.

Any towels or hive tools that have become smeared with foul honey should be boiled, and if the honey is extracted it should also be boiled before being fed back.

As I said at the beginning of this chapter, if colonies are kept strong one need have little fear of disease.

I might say in passing that the brown and the black bees are more prone to these diseases than the yellow races, which is another argument in favor of keeping only Italian bees in the apiary.

CHAPTER XIII

ENEMIES OF BEES

Lock the stable door before the horse gets out.—Vincent.

To say nothing of two-legged thieves, there are a number of enemies that constantly lie in wait to pounce upon the busy bees. Ants, bee-moths, birds, skunks, toads, wasps and even bears all have a liking for honey, and it is well to keep a constant lookout for them and, as far as possible, prevent their ravages.

Where thieving rascals rob apiaries, a resort to the law is the most effective method of stopping it; and if the identity of the thief is not known a sign hung in a conspicuous place offering a reward for information concerning the thieves will, in most cases, prove effective.

As with diseases of bees, so with enemies,—a little care will almost entirely remove their ravages. The ignorant beekeeper is perhaps the worst enemy with which the bees have to contend.

MICE.—Mice do harm only when they get into the hives or when they have access to empty combs that have been improperly stored for the winter.

Generally speaking, a strong colony will take care of any stray mouse that may be unfortunate enough to enter a hive, but there are times when a weak colony will not resist the entrance of mice in the late fall, and I know by experience that they can do considerable damage.

Even in the case of a weak colony, they can be

kept out if the entrance to the hive is not deeper than a quarter of an inch.

If surplus combs are not packed away in mouse-proof boxes, mice will get after them and completely riddle them.

SKUNKS.—In some localities these odoriferous little rascals are a positive nuisance, and have the

KILLING A SKUNK CAUGHT IN THE ACT OF CATCHING AND KILLING LIVE BEES

habit of approaching the hives at night; by scratching on the entrances of the hives they lure the bees out and enjoy a luscious repast. How these imps succeed in eating the bees without being stung is a mystery to me; but eat them they do, with little fear of stings.

The only way to get rid of these pests is to

set traps, properly baited, about the yard; and, after two or three are caught, the rest seem to become wise and give the apiary a wide berth.

ANTS.—Sometimes ants will make their formaries under a hive of bees and will be a nuisance to the beekeeper, but they can be gotten rid of by pouring a hot carbolic-acid solution on their nests.

BIRDS.—King-birds and other insectivorous birds will catch and eat a bee, and often a virgin queen

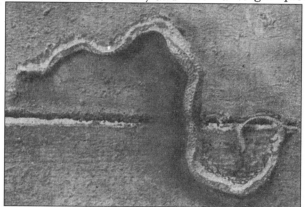

THIS SNAKE INTRUDED INTO A HIVE; THE BEES KILLED HIM, BUT BEING UNABLE TO REMOVE THE BODY, THEY GLUED IT TO THE BOTTOM BOARD WITH PROPOLIS

when on her matrimonial flight; but this is a chance that the beekeeper must take, and usually the damage done is inconsiderable.

WASPS.—Wasps and hornets will often capture and eat honey-bees, but as few are lost in this way little concern need be given them. In the South, notably in Florida, large numbers of virgin queens in flight are captured by the dragon flies with which that state abounds.

"THE PROOF OF THE PUDDING" — SUCH HONEY!

CHAPTER XIV

HONEY AS A FOOD

Eat honey because it is good.—Proverbs 24 : 13.

If the food value of honey were fully realized by our people it would oftener be found on our tables, not only during griddle-cake season but throughout the entire year. It is a known fact that honey is a predigested food, made so by the bees, and that it does not tax the digestive organs as do other sweets.

For centuries our forefathers had no other sweet save honey, but in recent times refined sugars have become so common that honey has been put to one side as a luxury.

When you realize that the possession of a few colonies of bees makes it possible to produce honey at a ridiculously low cost, you can see how important it is that folks should keep bees.

It is estimated that in the United States the consumption of sugar averages about eighty pounds per capita annually, while in the British Isles the average is much greater. This shows that the human system craves sweets of some kind; and from the standpoint of health it is better in many ways that this craving should be satisfied, at least in a large measure, with honey.

It is superior to sugar in many ways, having an aroma of its own, and when used in cakes and cookies will keep them moist and fresh much longer than sugar.

Honey is a better sweet than sugar, in that it does not tax the system to throw off a lot of useless material, as perhaps not more than one two-hundredth part of it is actual waste.

The children especially crave sweets to build up their rapidly developing bodies, and honey can be made to fill this need without the attendant digestive disturbances that so frequently follow free indulgence in candy and sugar.

Many of the large baking concerns use enormous quantities of honey in the making of honey cakes, and instances are known where they have kept fresh and sweet twelve years, requiring only to be placed in a damp place for a few days to restore them to their original condition.

For permission to print here the following recipes I am indebted to the courtesy of the A. I. Root Company. These recipes show but a few of the many uses to which honey may be put.

AIKIN'S HONEY COOKIES.—One teacupful extracted honey, one pint sour cream, scant teaspoonful soda, flavoring if desired, flour to make a soft dough.

FOWLS' HONEY COOKIES.—Three teaspoonfuls soda dissolved in two cupfuls warm honey, one cupful shortening containing salt, two teaspoonfuls ginger, one cupful hot water, flour sufficient to roll.

FOWLS' HONEY FRUIT CAKE.—One-half cupful butter, three-quarters cupful honey, one-third cupful apple jelly or boiled cider, two eggs well beaten, one teaspoonful soda, one teaspoonful each of cinnamon, cloves and nutmeg, one teacupful each of raisins and dried currants. Warm the butter, honey and apple jelly slightly, add the beaten eggs, then the soda dissolved in a little warm water; add spices and flour enough to make a stiff batter, then stir in the fruit and bake in a slow oven. Keep in a covered jar several weeks before using.

FOWLS' HONEY LAYER CAKE.—Two-thirds cupful butter, one cupful honey, three eggs beaten, one-half cupful milk. Cream the honey and butter together, then add the eggs and milk. Then add two cupfuls flour containing one and one-half teaspoonfuls baking powder previously stirred in. Then stir in flour to make a stiff batter. Bake in jelly-tins. When the cakes are cold, take finely flavored candied honey, and after creaming it spread between layers.

GINGER HONEY CAKE.—One cupful honey, one-half cupful butter or drippings, one tablespoonful boiled cider, in half a cupful of hot water (or one-half cupful sour milk will do instead). Warm these ingredients together, and then add one tablespoonful ginger and one teaspoonful soda sifted in with flour enough to make a soft batter. Bake in a flat pan.

HONEY APPLE BUTTER.—One gallon good cooking apples, one quart honey, one quart honey vinegar, one heaping teaspoonful ground cinnamon. Cook several hours, stirring often to prevent burning. If the vinegar is very strong, use part water.

HONEY CAKE OR COOKIES, WITHOUT SUGAR OR MOLASSES.—Two cupfuls honey, one cupful butter, four eggs (mix well), one cupful buttermilk (mix), one good quart flour, one level teaspoonful soda or saleratus. If it is too thin, stir in a little more flour. If too thin it will fall. It does not want to be as thin as sugar cake. I use very thick honey. Be sure to use the same cup for measuring. Be sure to mix the honey, eggs and butter well together. You can make it richer if you like by using clabbered cream instead of buttermilk. Bake in a rather slow oven, as it burns very easily. To make the cookies use a little more flour, so that they will roll out well without sticking to the board. Any kind of flavoring will do. I use ground orange peel mixed soft. It makes a very nice gingerbread.

HONEY CARAMELS.—One cupful extracted honey of best flavor, one cupful granulated sugar, three tablespoonfuls sweet cream or milk. Boil to "soft crack," or until it hardens when dropped into cold water, but not too brittle—just so it will form into a soft ball when taken in the fingers. Pour into a greased dish, stirring in a teaspoonful extract of vanilla just before taking off. Let it be one-half or three-quarters inch deep in the dish, and as it cools cut in squares and wrap each square in paraffine paper, such as grocers wrap butter in. To make chocolate caramels add to the foregoing one tablespoonful melted chocolate just before taking off the stove, stirring it in well. For chocolate caramels it is not so important that the honey be of best quality.

HONEY-DROP CAKES.—One cupful honey, one-half cupful sugar, one-half cupful butter or lard, one-half cupful sour milk, one egg, one-half tablespoonful soda, four cupfuls sifted flour.

HONEY FRUIT CAKE.—One and one-half cupfuls honey, two-thirds cupfuls butter, one-half cupful sweet milk, two eggs well beaten, three cupfuls flour, two teaspoonfuls baking powder, two cupfuls raisins, one teaspoonful each of cloves and cinnamon.

HONEY GEMS.—Two quarts flour, three tablespoonfuls melted lard, three-quarters pint honey, one-half pint molasses, four heaping tablespoonfuls brown sugar, one and one-half level tablespoonfuls soda, one level teaspoonful salt, one-third pint water, one-half teaspoonful extract vanilla.

HONEY GINGERSNAPS.—One pint honey, three-quarters pound butter, two teaspoonfuls ginger. Boil together a few minutes, and when nearly cold put in flour until it is stiff. Roll out thin, and bake quickly.

HONEY JUMBLES.—Two quarts flour, three tablespoonfuls melted lard, one pint honey, one-quarter pint molasses, one and one-half level tablespoonfuls soda, one level teaspoonful salt, one-quarter pint water, one-half teaspoonful vanilla.

These jumbles and the gems preceding are from recipes used by bakeries and confectioneries on a large scale, one firm in Wisconsin alone using ten tons of honey annually in their manufacture.

HONEY NUTCAKES.—Eight cupfuls sugar, two cupfuls honey, four cupfuls milk or water, one pound almonds, one pound English walnuts, three cents' worth each of candied lemon and orange peel, five cents' worth citron (the last three cut fine), two large tablespoonfuls soda, two teaspoonfuls cinnamon, two teaspoonfuls ground cloves. Put the milk, sugar and honey on the stove, to boil fifteen minutes; skim off the scum and take from the stove. Put in the nuts, spices and candied fruit. Stir in as much flour as can be done with a spoon. Set away to cool, then mix in the soda (don't make the dough too stiff). Cover up and let stand over night, then work in flour enough to make a stiff dough. Bake when you get ready. It is well to let it stand a few days, as it will not stick so badly. Roll out a little thicker than a common cooky, cut in any shape you like.

This recipe originated in Germany, is old and tried, and the cake will keep a year or more.

HONEY POPCORN BALLS.—Take one pint extracted honey; put it into an iron frying-pan, and boil until very thick; then stir in freshly popped corn, and when cool mold into balls. These will specially delight the children.

HONEY SHORTCAKE.—Three cupfuls flour, two teaspoonfuls baking powder, one teaspoonful salt, one-half cupful shortening, one and one-half cupfuls sweet milk. Roll quickly, and bake in a hot oven. When done, split the cake and spread the lower half thinly with butter, and the upper half with one-half pound of the best-flavored honey. (Candied honey is preferred. If too hard to spread well it should be slightly warmed or creamed with a knife.) Let it stand a few minutes, and the honey will melt gradually, and the flavor will permeate all through the cake. To be eaten with milk.

HONEY TEA CAKE.—One cupful honey, one-half cupful sour cream, two eggs, one-half cupful butter, two cupfuls flour, scant half teaspoonful soda, one tablespoonful cream of tartar. Bake thirty minutes in a moderate oven.

HOWELL HONEY CAKE.—(It is a hard cake). Take six pounds flour, three pounds honey, one and one-half pounds sugar, one and one-half pounds butter, six eggs, one-half ounce saleratus; ginger to your taste. Directions for mixing: Have the flour in a pan or tray. Pack a cavity in the center. Beat the honey and yolks of eggs together well. Beat the butter and sugar to cream, and put into the cavity in the flour; then add the honey and yolks of the eggs. Mix well with the hand, adding a little at a time, during the mixing, the half ounce saleratus dissolved in boiling water until it is all in. Add the ginger, and finally add the whites of the six eggs, well beaten. Mix well with the hand to a smooth dough. Divide the dough into seven equal parts, and roll out like gingerbread. Bake in ordinary square pans made for pies, from 10x14 tin. After putting into the pans mark off the top in one-half inch strips with something sharp. Bake an hour in a moderate oven. Be careful not to burn, but bake well. Dissolve sugar to glaze over top of cake. To keep the cake, stand on end in an oak tub, tin can or stone crock —crock is best. Stand the cakes up so the flat sides will not touch each other. Cover tight. Keep in a cool, dry place. Don't use until three months old at least. The cake improves with age, and will keep good as long as you will let it. I find any cake sweetened with honey does not dry out like sugar or molasses cake, and age improves or develops the honey flavor.

SOFT HONEY CAKE.—One cupful butter, two cupfuls honey, two eggs, one cupful sour milk, two teaspoonfuls soda, one teaspoonful ginger, one teaspoonful cinnamon, four cupfuls flour.

SUMMER HONEY DRINK.—One spoonful fruit juice and one spoonful honey in one-half glass water; stir in as much soda as will lie on a silver dime, and then stir in half as much tartaric acid and drink at once.

ABANDONED STREET CAR USED AS A BEEKEEPER'S WORK HOUSE

CHAPTER XV

HONEY AS A MEDICINE

Cæsar, when dining with his friend, Pollio, inquired how he had preserved himself to such a hale and hearty old age. Pollio replied: "Internally with honey, externally with oil."

If honey holds a place as a perfect food, it also is important as a medicine, and is largely used by wholesale drug houses in compounding many of the popular remedies on the market. Apart from being an agreeable medium in which to administer nauseous medicines, its value as a remedial specific in many disorders is recognized by the medical fraternity.

By its regular use the body is greatly benefited, especially in cases of colds and lung troubles. There is abundant testimony that chronic cases of bronchitis and asthma have been much helped by its use. Honey, tar and hoarhound are well known for colds.

In Germany, many old people make a delicious honey tea by mixing a couple of teaspoonfuls of extracted honey in a cupful of hot water; and they attribute their long lives to this alone.

Honey has a mild laxative effect, and is valuable in keeping the digestive organs in good condition. It has a specially beneficial effect upon delicate membranes, and is, therefore, good for stomach and intestinal troubles. Being predigested by the bees, it is immediately assimilated without fermentation, and becomes at once a build-er-up of the constitution.

In Denmark, honey is very popular with physicians in the treatment of chlorosis, based on the theory that the insufficiency of sweets in the system is responsible for the anæmic condition for which it is prescribed.

Nursing babes are cured of constipation by its use, and there is no period of life in which it is not of decided benefit.

To the A. I. Root Company I am indebted for permission to reprint the following hints and remedies that any one can practise or compound with honey:

COUGHS, COLDS, WHOOPING COUGH, ETC.—Fill a bell-metal kettle with hoarhound leaves and soft water, letting it boil until the liquor becomes strong —then strain through a muslin cloth, adding as much honey as desired—then cook it in the same kettle until the water evaporates, when the candy may be poured into shallow vessels and remain until needed, or pulled like molasses candy until white.

DR. KNEIPP'S HONEY SALVE.—This is recommended as an excellent dressing for sores and boils. Take equal parts of honey and flour, add a little water and stir it thoroughly. Don't make too thin. Then apply as usual.

HONEY AND CREAM FOR FRECKLES.—Have you tried a mixture of honey and cream—half and half —for freckles? Well, it's a good thing. If on the hands, wear gloves on going to bed.

HONEY CROUP REMEDY.—This is the best known to the medical profession, and is an infallible remedy in all cases of mucous and spasmodic croup:

Raw linseed oil, two ounces; tincture of blood root, two drachms; tincture of lobelia, two drachms; tincture of aconite, one-half drachm; honey, four ounces. Mix. Dose, one-half to one teaspoonful every fifteen to twenty minutes, according to the urgency of the case. It is also excellent in all throat and lung troubles originating from a cold. This is an excellent remedy in lung trouble: Make a strong decoction of hoarhound herb and sweeten with honey. Take a tablespoonful four or five times a day.

HONEY FOR DYSPEPSIA.—A young man who was troubled with dyspepsia was advised to try honey and graham gems for breakfast. He did so, and commenced to gain, and now enjoys as good health as the average man; and he does not take medicine, either. Honey is the only food taken into the stomach that leaves no residue; it requires no action of the stomach whatever to digest it, as it is merely absorbed and taken up into the system by the action of the blood. Honey is the natural foe to dyspepsia and indigestion, as well as a food for the human system.

HONEY AS A LAXATIVE.—In olden time the good effects of honey as a remedial agent were well known, but of late little use is made thereof. A great mistake, surely. Notably is honey valuable in constipation. Not as an immediate cure, like some medicines which momentarily give relief, only to leave the case worse than ever afterward, but by its persistent daily use, bringing about a healthy condition of the bowels, enabling them properly to perform their functions. Many suffer daily from an irritable condition, calling themselves nervous, and all that sort of thing, not realizing that constipation is at the root of the matter, and that a faithful daily use of honey fairly persisted in would restore cheerfulness of mind and a healthy body.

HONEY FOR OLD PEOPLE'S COUGHS.—Old people's coughs are as distinct as the coughs of children, and require remedies especially adapted to them. It is known by the constant tickling in the pit of the throat—just where the Adam's apple projects —and is caused by phlegm that accumulates there, which, owing to their weakened condition, they are unable to expectorate.

Take a fair-sized onion—a good, strong one— and let it simmer in a quart of honey for several hours, after which strain and take a teaspoonful frequently. It eases the cough wonderfully, though it may not cure.

HONEY FOR STOMACH COUGH.—All mothers know what a stomach cough is—caused by an irritation of that organ, frequently attended with indigestion. The child often "throws up" after coughing.

Dig down to the roots of a wild cherry tree, and peel off a handful of the bark, put it into a pint of water, and boil down to a teacupful. Put this tea into a quart of honey, and give a teaspoonful every hour or two. It is pleasant, and if the child should also have worms, which often happens, they are pretty apt to be disposed of, as they have no love for the wild-cherry flavor.

HONEY AND TAR COUGH CANDY.—Put a double handful of green hoarhound into two quarts of water, boil down to one quart; strain, and add to this tea two cupfuls of extracted honey and a tablespoonful each of lard and tar. Boil down to a candy, but not enough to make it brittle. Begin to eat this, increase from a piece the size of a pea to as much as can be relished. It is an excellent cough candy, and always gives relief in a short time.

HONEY AND TAR COUGH CURE.—Put one tablespoonful liquid tar into a shallow tin dish, and place it in boiling water until the tar is hot. To this add a pint of extracted honey, and stir well for half an hour, adding to it a level teaspoonful pulverized borax. Keep well corked in a bottle. Dose, one teaspoonful every one, two or three hours, according to severity of cough.

A SOLAR WAX-EXTRACTOR

Chapter XVI

BEESWAX

*How skillfully she builds her cell, how neat she spreads the
  wax,*
And labors hard to store it well, with the sweet food she makes.
—Watts' Divine Songs.

Even where bees are kept in limited numbers, there will accumulate in the course of a season quite a quantity of beeswax from old combs, cappings from extracting, etc. This wax, if properly rendered, besides forming a considerable source of revenue, is indispensable to the beekeeper, as nothing has been discovered that will take its place for the making of foundation.

WAX SCALES ON ABDOMEN OF WORKER BEE

The practise of most beekeepers is to melt it up and send it to the supply house, taking foundation in return; for it is a needless expense, to say nothing of the time required, to undertake to manufacture one's own foundation.

In the arts, as well as in the commercial field, beeswax is an important article, and the beekeeper will do well to save even the smallest particles and melt them up at his leisure.

The original method of rendering wax was by aid of the solar wax-extractor first used in California in 1862; and while still used to a certain extent by some, has been replaced by wax presses using steam or boiling water, for a larger percentage of wax can thereby be secured, especially in the case of old combs.

There are many presses made for this purpose, the best of which is the Hatch Gamil, which can be procured from the various supply houses.

The old combs are placed in sacks (feed sacks will do), boiled for a while and then put into the press under pressure.

A small cider-press answers very well and in principle is similar to the press referred to. Even a lard press may do good service, provided the old combs are properly boiled.

There will probably never be a surplus of beeswax on the market, as its use is essential in the sciences and arts, and no satisfactory substitute has yet been found.

I am amazed at the great loss of beeswax to some beekeepers who do not render it properly, as nearly twelve per cent. is often left in the slum gum which is thrown out as waste. This accounts for the many advertisements in bee journals where careful beekeepers seek to buy slum gum; they are able to extract a large amount of wax from it, even after the average man has thrown it aside as worthless.

The main thing to be followed in getting it all

is to steam and re-steam the mass of old combs, and to have a press with sufficient pressure to get it all as far as possible. Of course there will be left in the slum gum, even after careful render-

A GOOD WAX PRESS

ing, possibly four per cent. of wax, but this is far better than losing twelve per cent. of it.

Any man is foolish (to say nothing about being dishonest) to adulterate his wax in any way.

SENTINEL BEES

CHAPTER XVII

HONEY PLANTS

*"How doth the little busy bee improve each shining hour,
And gather honey all the day from every opening flower."*

There is no section of the country in which agricultural pursuits are carried on where bee-keeping can not be made very profitable. It is a mistake to suppose that the surrounding neighborhood must be a paradise of clover, buckwheat or basswood, for there are many unexpected sources even in unfavorable localities from which the busy bees will procure a good surplus for their owners.

The advantage of favorable location simply means that the beekeeper can profitably maintain a greater number of colonies in a given place. Where the bloom is not so abundant, to make bee-keeping a success the keeper will have to resort to the system of out apiaries, mentioned elsewhere in this book.

Very little if any honey is secured as a surplus from the early fruit bloom, as the bees use this mainly in raising bees to be ready for the clover flow, so that the clovers form, with basswood, the main sources of the early and white honey.

In some localities the sweet bush clover is a very important source of honey and should not be cut down as a weed, as it furnishes nectar for a considerable length of time.

The development of the red clover strain of bees now makes it possible to secure a large return from a plant which until recently was overlooked entirely by the bees, owing to their inability to reach the nectar in the deep corolla of the blossom.

Alsike, which is becoming more and more popular as a forage crop, is one of the best honey-producing plants in the world, and if mixed with timothy will overcome the objection raised to its cultivation, namely, that it does not stand up well. I know a number of beekeepers who find it profitable to supply the neighboring farmers with the seed free, as the returns in honey are an hundred-fold.

Goldenrod and the wild asters are heavy producers of a late fall flow, and are abundant in many parts of the country.

It would hardly be wise to raise special crops on the farm for honey, although this has been done with alsike, buckwheat, etc. The better way would be to supply the seed to nearby farmers; and where there are several beekeepers in the neighborhood, divide up the expense, as all will be benefited.

The list of flowers that are of value as honey and pollen-producing ones is long, and the following list taken from a United States Department of Agriculture bulletin will give you some idea of the many sources of supply. An effort has been made to indicate by the type the relative importance of the plants as pollen and honey producers; the larger the type the more important the plant.

NORTH AND NORTHEAST

NAME	[Above 40° N]	TIME OF BLOOM
Red or Soft Maple (*Acer rubrum*)	. . . . .	April.
Alders (*Alnus*)	. . . .	April.
Elm (*Ulmus*)	. . . .	April,

Willows (*Salix*) . . . . Apr.--May.

Dandelion (Taraxacum taraxacum= T. officinale
of Gray's Manual) . . . . Apr.–May.

Sugar, Rock, or *Hard Maple (Acer saccharum =*
A. saccharinum of Gray's Manual) . . . . Apr.–May.

Juneberry, or *Service Berry (Amelanchier cana-*
densis) . . . . May.

Wild Crab Apples (*Pyrus*) . . . . May.

Gooseberry and Currant (*Ribes*) . . . . . May.

Peach, Cherry, and Plum (*Prunus*) . . . . May.

Pear and Apple (*Pyrus*) . . . . May.

Huckleberries and Blueberries (*Gaylussacia* and
Vaccinium) . . . . May–June.

Common, Black, or Yellow Locust (*Robinia*
pseudacacia) . . . . May–June.

European Horse-chestnut (*Æsculus hippocasta-*
num) . . . . May–June.

Common Barberry (*Berberis vulgaris*) . . . . May–June.

Tulip Tree, or "Whitewood" (*Liriodendron*
tulipifera) . . . . May–June.

Grapevines (*Vitis*) . . . . May–June.

Rape (*Brassica napus*) . . . . May-June.

White Mustard and *Black Mustard* (*Brassica*
alba and *B. nigra*) . . . . June.

Raspberry (*Rubus*) . . . . June.

White Clover (*Trifolium repens*) . . . . June–July.

Alsike Clover (*Trifolium hybridum*) . . June–July.

Edible Chestnut (*Castanea dentata = C. sativa* var.
americana of Gray's Manual) . . . . . . June–July.

Alfalfa, or Lucern (*Medicago sativa*) . . . June–July.

Linden, or Basswood (*Tilia americana*) . July.

Smooth Sumac (*Rhus glabra*) . . . . July.

Buttonbush (*Cephalanthus occidentalis*) . . . . July.

MELILOT, BOKHARA, or SWEET Clover (*Melilotus*

 alba) . . . . July–Aug.

Indian Corn (*Zea mays*) . . . . July–Aug.

Melon, Cucumber, Squash, Pumpkin (*Citrullus*,

 Cucumis, and *Cucurbita*) . . . . July–Aug.

Fireweed (*Erechthites hieracifolia*) . . . . . . July–Sept.

Chicory (*Cichorium intybus*) . . . . July–Sept.

KNOTWEEDS (*Polygonum*, especially *P. pennsyl-*

 vanicum and *P. persicaria*) . . . . . . Aug.–Sept.

BUCKWHEAT (*Fagopyrum fagopyrum = F.*

 esculentum of Gray's Manual) . . . . . . Aug.–Sept.

Indian Currant, or Coral Berry (*Symphoricarpos*

 symphoricarpos = S. vulgaris of Gray's

 Manual) . . . . Aug.–Sept.

GREAT WILLOW-HERB (*Epilobium angustifolium*)

Aug.–Sept.

Thoroughwort, or Boneset (*Eupatorium perfoliatum*)

Aug.–Sept.

Bur Marigolds (*Bidens*, especially SPANISH

 NEEDLES, *Bidens bipinnata*) . . . . . Aug.–Oct.

Wild Asters (*Aster*) . . . . Aug.–Oct.

GOLDENRODS (*Solidago*) . . . . Aug.–Oct.

MIDDLE SECTION

[Between 35° and 40° N]

Redbud (*Cercis canadensis*) . . . . Mar. –Apr.

Alder (*Alnus rugosa = A. serrulata* of Gray's

 Manual) . . . . Mar. –Apr.

Red or Soft Maple (*Acer rubrum*) . . . . . . Mar. –Apr.

Elm (*Ulmus*) . . . . Mar. –Apr.

Willows (*Salix*) . . . . Mar.–May.

Dandelion (*Taraxacum taraxacum* = *T. officinale*
of Gray's Manual) . . . . Apr.–May.
Apricot (*Prunus armeniaca*) . . . . Apr.–May.
Juneberry (*Amelanchier canadensis*) . . . . . Apr.–May.
Wild Crab Apples (*Pyrus*) . . . . Apr.–May.
Gooseberry and *Currant* (*Ribes*) . . . . . . Apr.–May.
Rhododendrons (*Rhododendron*) . . . . . . Apr.–May.
Peach, Cherry, and *Plum* (*Prunus*) . . . . Apr.–May.
Pear and *Apple* (*Pyrus*) . . . . Apr.–May.
CRIMSON CLOVER (*Trifolium incarnatum*) . . Apr.–May.
Huckleberries and Blueberries (*Gaylussacia* and
Vaccinium) . . . . May.
American Holly (*Ilex opaca*) . . . . May.
Black Gum, Sour Gum, Tupelo or *Pepperidge*
(*Nyssa aquatica* = *N. sylvatica* of Gray's
Manual) . . . . May.
Manzanitas (*Arctostaphylos*) (California) . . . May.
COMMON, BLACK, or YELLOW LOCUST (*Robinia
pseudacacia*) . . . . May.
Barberry (*Berberis canadensis*) . . . . . . May.
TULIP TREE, or "POPLAR" (*Liriodendron
tulipifera*) . . . . May.
Mountain Laurel (*Kalmia latifolia*) . . . . May–June.
Grapevines (*Vitis*) . . . . May–June.
Persimmon (*Diospyros virginiana*) . . . . May–June.
White Clover (*Trifolium repens*) . . . . . May–June.
Alsike Clover (*Trifolium hybridum*) . . . . May–June.
RASPBERRY (*Rubus*) . . . . May–June.
COWPEA (*Vigna sinensis*) . . . . May–Aug.
EDIBLE CHESTNUT (*Castanea dentata* = *C. sativa*
var. *americana* of Gray's Manual) . . . . June.

Chinquapin (*Castanea pumila*) . . . . . . June.

Catalpas, or Indian Bean Trees (*Catalpa*) . . . June.

MAGNOLIA, or SWEET BAY (*Magnolia Glauca*) . June.

LINDEN, or "LINN" (*Tilia americana*) . . . June.

SOURWOOD, or SORRELL TREE (*Oxyden-drum arboreum*) . . . . June–July.

Oxeye Daisy, or *Whiteweed* (*Chrysanthemum leucanthemum*) . . . . June–July.

Smooth Sumac (*Rhus glabra*) . . . . July.

Buttonbush (*Cephalanthus occidentalis*) . . . . July.

CLEOME, or "ROCKY MOUNTAIN BEE PLANT" (*Cleome serrulata = C. integrifolia* of Gray's Manual) (West) . . . . July–Aug.

ALFALFA (*Medicago sativa*) (West) . . . . July–Aug.

MELILOT, BOKHARA, or SWEET CLOVER (*Melilotus alba*) . . . . July–Aug.

Indian Corn (*Zea mays*) . . . . July–Aug.

Cucumber, Melon, Squash, Pumpkin (*Cucumis*, *Citrullus*, and *Cucurbita*) . . . . July–Aug.

Knotweeds (*Polygonum*, especially *P. pennsylvanicum* and *P. persicaria*) . . . . . . July–Sept.

Buckwheat (*Fagopyrum fagopyrum = F. esculentum* of Gray's Manual) . . . . Aug.–Sept.

Wild Asters (*Aster*, especially HEATH-LIKE ASTER, *Aster ericoides*) . . . . Aug.–Oct.

Thoroughwort, or Boneset (*Eupatorium perfoliatum*) . . . . Aug.–Oct.

Bur Marigolds (*Bidens*, especially SPANISH NEEDLES, *Bidens bipinnata*) . . . . . . Aug.–Oct.

Goldenrods (*Solidago*) . . . . Aug.–Oct.

SOUTH
[Below 35° N.]

Redbud (*Cercis canadensis*) . . . . Feb. – Mar.

Alder (*Alnus rugosa = A. serrulata* of Gray's
Manual) . . . . Feb. – Mar.

Red or Soft Maple (*Acer rubrum*) . . . . . . Feb. – Mar.

Elm (*Ulmus*) . . . . Feb. – Mar.

Willows (*Salix*) . . . . Feb. – Mar.

Dandelion (*Taraxacum taraxacum = T. officinale*
of Gray's Manual) . . . . Feb. – Mar.

Apricot (*Prunus armeniaca*) . . . . Feb. – Mar.

Carolina Cherry, or Laurel Cherry (*Prunus caro-
liniana*) . . . . March.

Juneberry (*Amelanchier canadensis*) . . . . . March.

ORANGE and *Lemon* (*Citrus*) . . . . Mar. – Apr.

Cottonwoods or *Poplars* (*Populus*) . . . . . Mar. – Apr.

TITI (*Cliftonia ligustrina*) (Florida and southern
Georgia, westward) . . . . Mar. – Apr.

Gooseberry and Currant (*Ribes*) . . . . . . Mar. – Apr.

Peach, *Cherry*, and *Plum* (*Prunus*) . . . . . Mar. – Apr.

Pear and Apple (*Pyrus*) . . . . Mar. – Apr.

Huckleberries and Blueberries (*Gaylussacia* and
Vaccinium) . . . . April.

Crimson Clover (*Trifolium incarnatum*) . . . April.

BLACK GUM, SOUR GUM, TUPELO, or PEPPERIDGE
(*Nyssa aquatica = N. sylvatica* of Gray's
Manual) . . . . April.

BALL, or BLACK SAGE (*Ramona stachyoides, R.
palmeri*, etc. *= Audibertia stachyoides*, etc.,
of the Botany of California) (California) . . April.

GALLBERRY, or HOLLY (*Ilex glabra*) . . . . . Apr. – May.

Manzanitas (*Arctostaphylos*) (California) . . . . Apr. – May.

Acacias (*Acacia*) . . . . Apr.–May.

Common, *Black*, or *Yellow Locust* (*Robinia pseudacacia*) . . . . Apr.–May.

Persimmon (*Diospyros virginiana*) . . . . . Apr.–May.

EDIBLE CHESTNUT (*Castanea dentata = C. sativa*
var. *americana* of Gray's Manual) . . . . Apr –May.

Chinquapin (*Castanea pumila*) . . . . . . Apr.–May.

Catalpas (*Catalpa*) . . . . Apr.–May.

MAGNOLIAS (*Magnolia*) . . . . Apr.–May.

Rhododendrons, Rosebays, Azaleas (*Rhododendron*) . . . . Apr.–May–June.

MESQUITE (*Prosopis juliflora*) (Texas and westward) . . . . Apr –July.

Cowpea (*Vigna sinensis*) . . . . Apr.–Aug.

TULIP TREE, or "POPLAR," (*Liriodendron
tulipifera*) . . . . May.

Mountain Laurel (*Kalmia latifolia*) . . . . . May.

Grapevines (*Vitis*) . . . . May.

Raspberry (*Rubus*) . . . . May.

China Berry, China Tree, or Pride of India
(*Melia azedarach*) . . . . May.

WHITE SAGE (*Ramona polystachya = Audibertia polystachya* of the Botany California)
(California) . . . . May–June.

HORSEMINT (*Monarda citriodora*) . . . . May– July.

SOURWOOD, or SORRELL TREE (*Oxyden-
drum arboreum*) . . . . May–June.

SAW PALMETTO (*Serenoa serrulata*) (coasts
of Georgia and Florida) . . . . May–June.

BANANA (*Musa sapientum*) . . . . May–Sept.

LINDEN, or "LINN" (*Tilia americana*) . . . June.

Red Bay (*Persea borbonia = P. carolinensis* of
    Gray's Manual) . . . . June.
Indian Corn (*Zea mays*) . . . . June–July.
Cucumber, Melon, Squash, Pumpkin (*Cucumis,*
    Citrullus, and *Cucurbita*) . . . . June–July.
CABBAGE PALMETTO (*Sabal palmetto*), coasts
    of South Carolina, Georgia, and Florida) . . June–July.
BLACK MANGROVE (*Avicennia tomentosa*
    and *A. oblongifolia*) (Florida) . . . . . June–July.
ALFALFA (*Medicago sativa*) . . . . June–Aug.
MELILOT, BOKHARA, or SWEET CLOVER (*Melilotus*
    alba) . . . . June–Aug.
COTTON (*Gossypium herbaceum*) . . . . . . June–Aug.
WILD PENNYROYAL (*Hedeoma pulegioides*) . . June–Sept.
BLUE GUM and RED GUM (*Eucalyptus globulus*
    and *E. rostrata*) (California) . . . . . July–Oct.
WILD BUCKWHEAT (*Eriogonum fasciculatum*)
    (California) . . . . Aug.–Sept.
Japan or *Bush Clover* (*Lespedeza striata*) . . . Aug.–Sept.
Bur Marigolds (*Bidens,* especially *Spanish Needles,*
    Bidens bipinnata) . . . . Aug.–frost.
Wild Asters (*Aster,* especially HEATH - LIKE
    ASTER, *Aster ericoides*) . . . . Aug.–frost.
Goldenrods (*Solidago*) . . . . Aug.–frost.

ROOF-TOP BEEKEEPING IN NEW YORK CITY

Chapter XVIII

MARKETING HONEY

"A prophet is not without honor save in his own country"
does not apply to the beekeeper who produces a choice article.

The question of disposing of the season's product is an important one, and upon its proper handling will hinge the question of profit. Except in the case of the honey producer whose product runs up into the tons and who for this reason will find it more convenient to ship his honey direct to commission merchants in the large cities, I should say by all means sell your honey to the local trade, and if a really first-class grade of well-ripened honey is produced, no trouble will be experienced in disposing of it at good prices.

Many grocers in nearby towns will be glad to handle it on consignment, and where this is not possible, a house-to-house canvass can be made with satisfactory results.

Comb-honey will of course be sold by the section; but in the case of extracted honey, an attractive bottle will be necessary. For this purpose I know of nothing better than the ordinary quart glass jar, as the jar is useful to the purchaser even after the honey is used.

A nice white label with photo-reproduction of the apiary on it, with owner's name and address, and a statement that it is pure honey, will completely fill the requirements.

It will be necessary to see that the extracted

honey is well strained and clear; heating it to, say, 145°, and keeping it there for three hours before bottling, will prevent its granulation for a considerable period. In heating it be very careful not to get it too hot, as that would spoil the delicate aroma of the honey.

Much depends upon the individual in the matter of selling honey to the local trade, and no rule can be laid down for everybody.

I know of one beekeeper whose method is to go to nearby towns with a large can of extracted honey. He rings the door-bell and when the lady of the house appears he politely informs her that he is giving away a sample of fine, pure honey, and if she will kindly bring a dish he will be pleased to have her try some. The offer is seldom refused, and a half-teacupful of honey is left at each home.

The following day, or later the same day, he returns and inquires if the honey was good, and if so would be pleased to take an order for a quart, gallon, or as much as will be desired; and he nearly always makes a sale. The honey is delivered in a day or two with his address on the package, and it is astonishing how many subsequent orders come to him by mail.

Another beekeeper goes among the offices of the larger towns, and has a little frame of honey and live bees in a glass-sided case with wire screen on top; and in this way he attracts attention and disposes of thousands of pounds every season, as many people appreciate the novelty of buying honey for the family direct from the beekeeper.

Still others bottle their honey in one-pound jars, two dozen in a case, and supply the groceries; and experience no difficulty in getting about $2.25 per dozen jars for them, as they retail for twenty-five cents per pound jar.

If, however, you feel that you are no salesman, then send your honey in bulk to the commission merchant in the cities, packing your comb-honey in cases (furnished by the supply houses) having corrugated paper in the bottom of each case to prevent breakage of combs.

When selling at wholesale, the best package is the regular sixty-pound can for extracted honey, two of which come in a case; the honey is more easily handled by the buyer when in this shape. As these cans are somewhat expensive, many producers of extracted honey pack their honey in the little kegs holding one hundred and sixty pounds, which were mentioned in the forepart of this book.

Under no circumstances pack your honey in old cans or kegs unless they have been thoroughly cleansed, or you will have a lot of fermented honey to take back; and see that none of the packages leak before shipping.

HIVES IN ORDERLY ROWS WITH GRASS AND WEEDS KEPT DOWN

Chapter XIX

BEEKEEPERS' CALENDAR

A time and place for everything, and everything in its time and place, spell success.—Martha.

Attention to little details at the proper season is what makes the successful beekeeper, and the reader will do well to make a study of this chapter.

January.—At this season of the year the bees are in a state of repose, and having been properly prepared in the fall as regards feeding and strengthening, will require no attention other than to look after the temperature and ventilation,—if wintered in the cellar.

From now on much can be done in the matter of ordering supplies and putting them together, so that this work will be out of the way when the rush comes in the spring and summer.

February.—While it is true that in some of the southern sections of the country bees begin to gather pollen and honey during this month, yet, speaking for the North, the same conditions prevail as during January.

March.—With the coming of March things begin to assume a different aspect, as the bees begin to bestir themselves, having wakened from their winter's sleep and begun work on the maples.

This is the month for removing the colonies from their winter cellars, and the late afternoon of a clear day is the best time to do it.

If any colonies have died, put their combs away from robber bees; and if any of such combs are heavy with honey, they may be used to slip into the brood nest of any that are short of stores.

APRIL.—As the bees are now gathering much pollen and some honey, they are rearing considerable brood, and if they have not sufficient food

AN APIARY SURROUNDED BY FRUIT ORCHARDS

it will mean a light honey surplus later on, as there will not be a sufficient working force to gather it.

For this reason it is a decided advantage to give them stimulative feed, using the jar with perforated cap referred to in the chapter on Spring Manipulation; there is no better all-around feeder in existence, and any one can quickly make one.

Feed about a half-pint of syrup every night for a month, and the results accomplished in the increased strength of the colonies fed will be astonishing.

If any colony is queenless, remedy it at once by buying a new queen, or else by giving it a frame of young eggs from a strong colony, for this is a condition that should not be continued.

MAY.—Colonies are building up rapidly now, and, unless measures are taken to prevent it, swarms may be looked for from the strong ones the latter part of the month.

Don't be in a hurry to put on the supers; better wait a while, unless the condition of the colonies demands it to prevent swarming.

JUNE.—While some swarms will emerge during May, yet it can be said without contradiction that this is the great swarming month of the year.

Put on the supers, as many as are needed, for the clover flow will soon begin and happy is the beekeeper who has an abundance of such.

Enlarge the entrances of the hives, cut out all queen cells and give added storage room as indicated, and swarming can be kept within reasonable bounds.

JULY.—About the middle or last of this month the clover flow will be over, and it is best to remove the surplus as soon as sealed, so as to keep it separate from the dark, fall honey that begins to come in during August.

With the end of the early flow look out for robbers, as these bees are alert to pounce down upon any hive that is opened even for a very few seconds.

AUGUST.—For the majority of beekeepers, it can be safely said that there is little or no honey gathered by the bees until the goldenrod, asters or other fall flowers come on.

In localities where buckwheat is extensively grown, August is the best season of the year for the beekeeper, as his early expectation of this late flow has permitted him to strengthen his weak colonies on the early flows so as to have them in prime condition now for this almost unfailing crop.

BEEKEEPING IS A NOBLE PROFESSION

It is the practise of progressive beekeepers to requeen all of their colonies during August, for by so doing each colony comes to the following spring headed by a vigorous layer, and the chances of swarming are greatly lessened by having young queens at the head of each colony.

SEPTEMBER.—In some sections of the country this is a busy month for the bees, especially if there is much fall bloom.

The latter part of this month is the time to feed up colonies for the winter, as they will then take up feed much better than later on, although it is the practise of many beekeepers to wait until the fall flow is entirely over in October.

OCTOBER.—Very little honey is now being gathered in the North, and for that reason the extracting and comb supers had better be taken off; so that the bees can store, what little may be gathered, in the brood nest for winter.

After extracting, be careful to put the combs in a safe place from the bee-moth, also from mice; and until freezing weather sets in it will be wise to examine them occasionally.

NOVEMBER.—Prepare your colonies for wintering. If wintered outdoors, put on their " winter overcoats."

The last of the month remove them to the cellar, if wintered indoors; doing so late in the afternoon.

DECEMBER.—If you haven't disposed of your honey crop before now, make haste to do so,— especially before the holidays. Read the chapter on Marketing Honey, and then get busy at once.

Look over the bee journals, evenings, and plan for a better year. Also figure up profit or loss and know just how you stand; don't guess at things.

Concluding words: Solomon has said, " Of making many books there is no end; and much study is a weariness of the flesh." Whether the perusal of this brief treatise has proved wearisome to the reader I do not know; but one thing I do know,—its writing has been an unalloyed pleasure.

I could have made it many times larger, but I question whether it would have been better. The plans and devices I have urged are the result of many years of experiment, and many thousands of miles of travel among the leading beekeepers of the country.

Everything I have urged has stood the test of time, and nothing that has been discarded has been placed before you.

Beekeeping is a noble profession, and it is my sincere desire, as well as that of Tim, Martha and my beloved wife, that farmers and gardeners everywhere may in the years to come be more and more successful in apiculture.

If this purpose is fulfilled, I have not toiled in vain.

INDEX

The Biggle Horse Book
by Jacob Biggle

"People ought to try to make their horses happy," wrote Jacob Biggle's wife, Harriet, in *The Biggle Horse Book* in 1894. "A happy, cheerful horse will do more work and live longer, and thus be more profitable to its owner, than one whose temper is kept constantly ruffled, whose disposition is soured by ill-usage, and whose peace of mind is often disturbed by the crack of the whip, the hoarse voice of the driver, the strain of overwork, the discomfort of a hard bed, or the pangs of hunger and thirst." When it comes to the treatment of animals—especially the horse—the Biggles were ahead of their time.

Folksy and informative, this manual offers timeless tips on the effective and humane treatment and training of horses and detailed descriptions of all the major breeds. Practical horsemen and veterinarians of the era contributed their wisdom and insight, and their maxims on owning, riding, and working with horses will provide endless hours of entertainment.

$9.95 Hardcover • ISBN 978-1-62636-145-4

The Biggle Poultry Book
by Jacob Biggle

When Jacob Biggle first published his book on the management of poultry, there were more than three hundred million chickens and thirty million other domesticated fowl in the United States. Today, the trend continues with thousands if not millions of chickens and other fowl being raised in suburban and urban backyards across America. Biggle's aim was to "help farmers and villagers conduct the poultry business with pleasure and profit."

Written for the practical farmer who raises poultry and eggs for market, *The Biggle Poultry Book* will also appeal to collectors of farm ephemera and anyone else who is nostalgic for a simpler way of doing things. Illustrated with sixteen color plates by Louis P. Graham, and hundreds of black-and-white photographs and illustrations throughout, *The Biggle Poultry Book* is as beautiful as it is useful and a treasure for the home library.

$9.95 Hardcover • ISBN 978-1-62636-147-8

The Biggle Swine Book
by Jacob Biggle

When Jacob Biggle first published *The Biggle Swine Book* in 1898, hog husbandry was undergoing major changes. New feeding methods had come into vogue, new breeds of hogs had been developed, and significant progress had been made in curbing swine-borne epidemics. Even the public perception of pigs as filthy creatures wallowing up to their knees in mud had brightened, and pigs were accorded a modicum of respect. But with the onset of railroad development across the United States, the backyard pig farmer started losing ground to slaughterhouses and large processing plants.

Illustrated with photographs, engravings, and line drawings throughout of all things pig-related, this book is a glimpse into a bygone era when sows and their litters had a place on every farm, and people knew exactly where their bacon came from.

$9.95 Hardcover • ISBN 978-1-62636-148-5